tITLE: **tHE bOOK oF p**
vOLUME fOUR

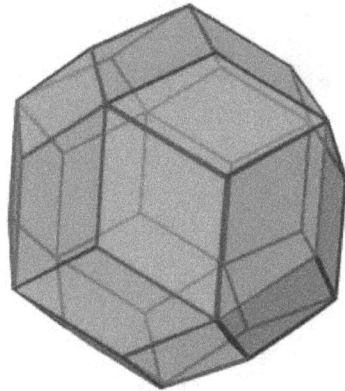

sUB-Title: **tHE 108 cODES:**
tHE lINEAR pHI cODE 1

eDITION 1

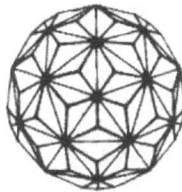

aUTHOR: **jAIN 108**

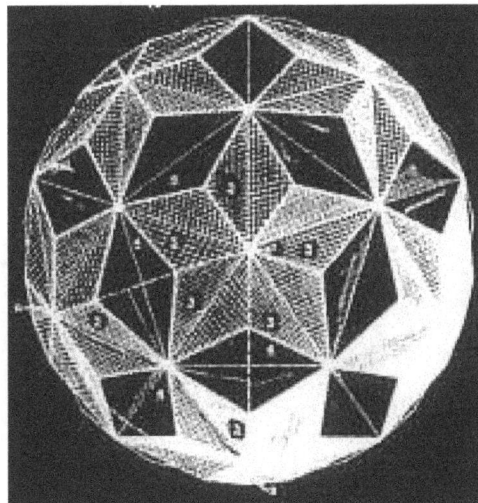

yEAR: 2010

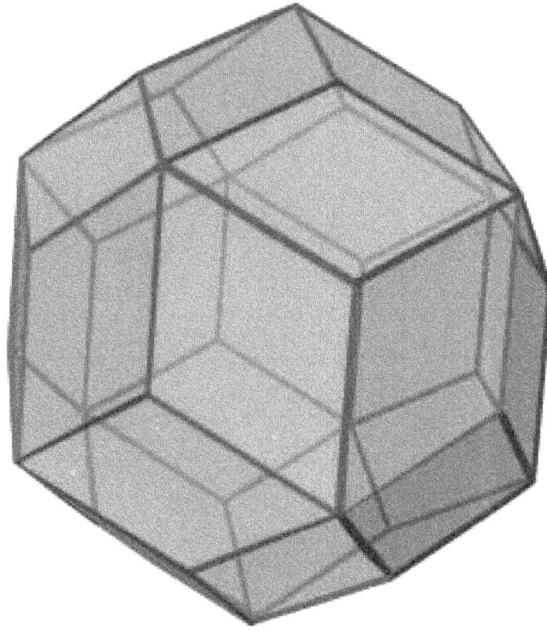

(Golden Rhombic TriaContaHedron Dreaming)

JAIN 108

PHI CODE 108

tITLE:

(tHE bOOK oF pHI vOLUME fOUR)

sUB-tITLE:

tHE 108 cODES: tHE lINEAR pHI cODE 1

edition 1

aUTHOR:

jAIN 108

yEAR:

2010

iSBN: 978-0-9757484-3-5

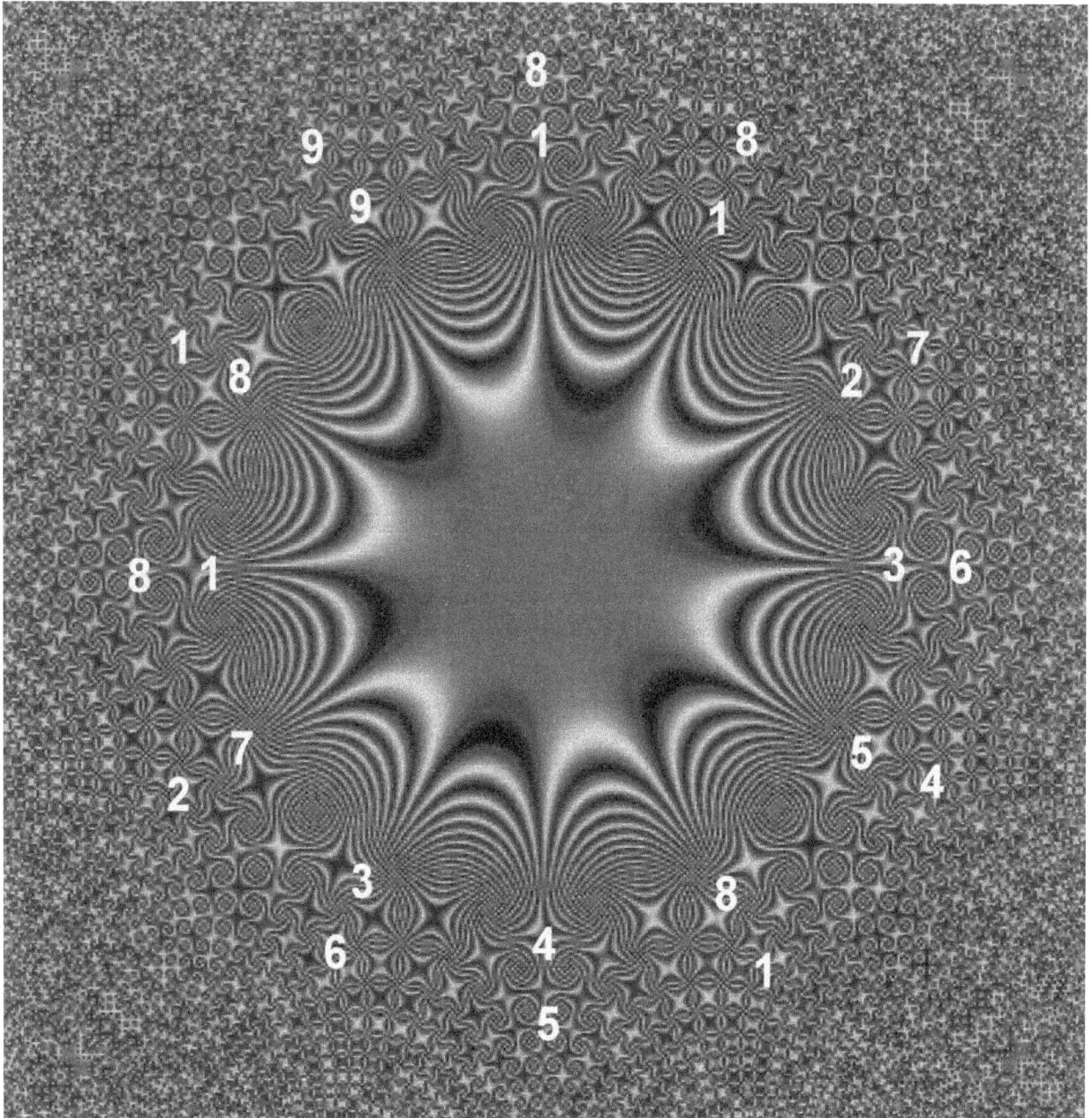

(Artwork by Mathematica program, representing a complex number.
I have inserted the 12 Pairs of 9 shown in white around the 12 sided pattern).

Intricate patterns can be found based on simple math rules combined with a complex number x + iy, where **i = the square root of minus one**.
(Even though we don't understand what is the square root of minus one, without it we could not send rockets or satellites to outer space or build complex computers).
image courtesy of:
http://demonstrations.wolfram.com/PatternsFromMathRulesUsingComplexNumbers/

the
book
of phi.
vol 4

phi code
108

(Original ovular border image taken from The Illustrated Encyclopedia of
"Flags and Heraldry" by Steven Slater and Alfred Znamierowski)

CONTENTS

"Blue Apples" by **Lily Moses**, 2010,
(sketch of the completed painting)

"Wherever there is number, there is beauty."
Proclus (410-485 A.D.)

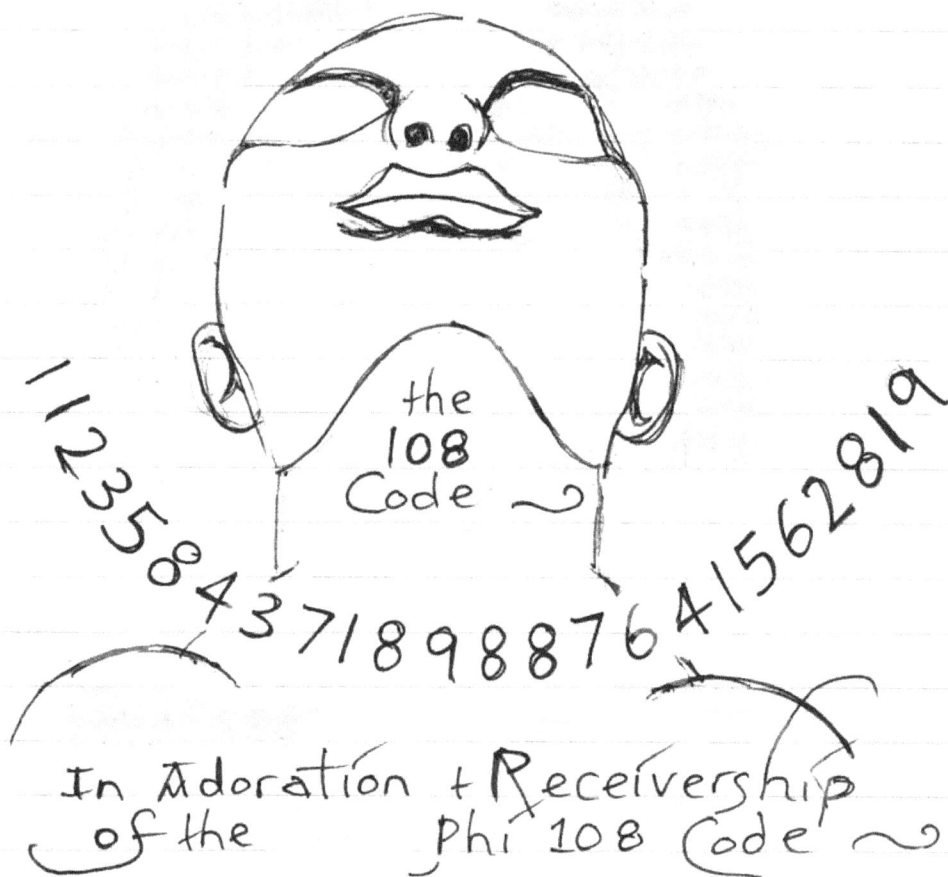

the
108
Code

1123584371898876415628 19

In Adoration + Receivership
of the   Phi 108 Code

(ART BY JAIN 2009)

INTRODUCTION

THE 108 PHI CODE TRANSMISSIONS

BY JAIN 108

by Jain, 09-09-09, Mullumbimby Creek, far north NSW

FREQUENTLY ASKED QUESTIONS

QUESTION 1:

Why did you call these phi codes 1 and 2, instead of some fancier names?

ANSWER:

Originally, I had two names but am not using them:
Phi Code 1 was known as the **GENERATOR**, as the compression of the fibonacci numbers generates the Phi Ratio of
1:1.618033988... on a linear, mathematical level.

Phi Code 2 was known as the **ELEVATOR**, derived from the Powers of Phi. Now that it has been generated, it allows or permits multi-dimensional access from the atomic to the galactic, a veritable Jacob's Ladder, the meaning of the Caduceus winged staff of the ancient medical order... It is really the true meaning of the Stairway To Heaven!

Either way, we have to give them names, and these names will change, but the Phi Codes 1 and 2 are a forever science, a mathematical fact of **fixed timeless design**.

QUESTION 2:
Here is an email I received from **Leonardo Fibonacci**.

re: subject matter: Phi Code = 108 + 9 eternal
Re: question about info on your webpage
Leonardo Fibonacci wrote:

"Hi Jain id been reading your web page and i found this math error, the sum of the numbers on the two series of twelve digits is 117 and not 108 as appears below, there must be a mistake somewhere, i guess is a printing error, not calculus. The difference is 9: (117-108=9).
If im wrong please let me know why.
Thanks.
If you take the first 12 digits and add them to the second twelve digits and apply numeric reduction to the result, you find that they all have a value of 9.

1st 12 Numbers   1 1 2 3 5 8 4 3 7 1 8 **9**
2nd 12 Numbers   8 8 7 6 4 1 5 6 2 8 1 **9**

The 12 Complementary Pairs of 9 in your stunning Phi Code certainly does express a divine eternal mathematics, what you call "The Galactic Bases of 9 & 12". These 12 pairs of 9 do indeed have a sum of 108 but if you include that last pair of the two nines or Double 9 anomaly seen at the end of the sequence, you will see it does all add up to **117**.

ps 1: Are the Indians praising you for this lost Vedic 108 Knowledge, or are they jealous or threatened that a westerner like you beat them to the press about it all.

Ps 2: Is this why no-one knows about Digital Compression or what we demeaningly call "Numerology" because its a doorway into the Mysteries?"

Yours:
Leonardo Fibonacci.

ANSWER:
My Response on the 10-3-2008:

Hi **Leonardo 1.618**
(email: f1.6180339887@gmail.com)

You are correct to observe this,
I have written two previous books about this very dilemma,
how to read this compressed data,
so to solve this briefly, let us say that the pattern is as you summed:
(108 + 9) where the floating 9 acts as a bridge or a double bond etc
so let say that the eternal repetition is thus:
(108 + 9) + (108 + 9) + (108 + 9) + (108 + 9) + (108 + 9) +...

There are many reasons why 108 is an anointed number.
I am currently writing 3 books about this whole topic now...
known as THE BOOK OF PHI volumes 3, 4 and 5.
So stay in touch, and thanks for your queries.

Regarding the controversial topics like the Number 13 and the Pentacle and Swaztika etc, I wrote a Newsletter Article on The **Demonization of Numerology**, posted currently on my website.

Is your real name Leonardo Fibonacci?
I was born in Sydney, Australia, therefore a westerner, but both my parents are full-blood Phoenician, therefore I am a total Arab, but my Soul is Indian! I have never had a Teacher except for Anahat (Heart) Guru, and am Self-Realized, giving Genuine Gratitude each day to extend my Latitude and raise my Altitude.

Regards Jain 108

(108 is my surname!
I have actually become a Number!)

JAIN F.R.E.E.D.O.M.S. trading as JAIN MATHEMAGICS
Email   jain@jainmathemagics.com
Web   www.jainmathemagics.com
Address pob 729, Mullumbimby, NSW, 2482. Australia

NOTES:

We are so marinated in the culture of drab education with lifeless lessons and dysfunctional mathematics. These 2 Phi Codes will re-wire the brain and join the neo-cortices to stimulate whole-brain holographic learning. They are therefore a celebration of The Truth of Mathematics and will take us to the next Threshold.

This book is a comprehensive anatomy of ancient mathematical structures, in the same guise as DNA's **numerical architecture**.

I knew I was a **Mathematical Troubadour** when I found the derivation of the word "Troubadour": Knowing that "Or" is Gold, thus in the French, it means literally "to find lost gold"
ie: a treasure.
At first I did not know whether I was a **numerical nomad**, or a **mathematical monk**, or an **arhythmophile**.
This **arhythmophile** is a specialist in **the living curvation** of seashells and the curvature modelling of **turtle mathematics**.

Ramtha's 4 Teachings:
1 – U R God
2 – Consciousness and Energy Create Reality
3 – We are here to make Known the Unknown
4 – The Challenge is to Conquer YourSelf

"Modern maths is like cerebral junk food, there's nothing vital or organic about it".
Quoted by:
$$(j.a^i)^n = \{ma + th^e \times [p(h)^i]\sim \}$$

$$\frac{\qquad}{ma_gic^s} \qquad \frac{\qquad}{1.o8^\wedge\#}$$

Aspire 2 Inspire b4 U Expire

Jain's knowledge of the Suppressed Mathematics for the **Tweener Market**, (aged 8 to 12)
is really a discourse about: "**Pneuma**" :
the Greek word for "Spirit" as non-gendered.

Mathemagics and Rapid Mental Calculation are a form of "**Kung-Fu**" when viewed by its original meaning or definition:
"any skill cultivated through long and hard work"...

The old maths we learnt at school still resides in a particular "**Blind Spot**" that you can't see. By continually teaching this old stuff we are staying stuck in our old ways. Its like we are blind. We need to move forward and open their eyes to the exciting and consciousness-changing shapes like the **30 Golden Faces** of the **Rhombic Triacontahedron** or **EarthStar** that magical polyhedral form (the same image that is on page 1) that is closest to the sphere and maps the topology of the world-map. This is the ultimate form that contains within its 32 vertices the 5 Platonic Solids. The outer sphere that encompasses this form is in the phi ratio 1:1.618 in regard to the lengths of the short and long axes. The dual of the Cuboctahedron (rhombic dodeca) is a similar form with 12 rhombi (of garnet crystal) but is based on the Root of 2 (1.4142...), my quest was to find a similar shape that is based all on the Golden Phi Mean of 1:1.618033... (that contains Root 5) and here it is, I found it after 30 years, it was looking at me all the time.

Becker-Hagens EarthStar®

World Map based on the Rhombic TriaContaHedron, that is, has 30 Golden Rhombi. Each phi diamond is quadrated to make a form of 120 right-angled triangles adored by Buckminster Fuller.

When we collectively realize that we are actually One Group, One Being, One Cell, as in the sense that the Earth is a cell of this Universe, then there can not exist any form of Plurality, as in stating "We Are". There is no "We Are". It can only be "We Is" as a Group Conscious Individual or an Individual Conscious Group. Not We Are, but rather WE IS: Pronounced: wizz!

In search of mathematical truth, I came across:
"Y gwir erbyn y byd"
which translates from the old Welsh language to mean:
"The Truth against the World"
I found this phrase to reference Boudicca's impassioned battle cry, She was a first century warrior queen who fought for the preservation of the Celtic race.
(mentioned in "The Expected One" by Kathleen McGowan, p431).

You are not sitting here reading this book to learn about mathematics. My path, my purpose is to provoke you, shock you to wake up and realize that most of the stuff you learned at school or uni were half-truths. By perusing through this book, you will see that indeed there is a glorious and galactic mathematics where Nature's anointed Phi Ratio (1:1.618033...) has a rhythmic dance, that Pi is married to Phi and the value you think it is equal to is a lie, to keep you dumbed-down, disharmonic. The Queen of England wears the Prime Number Cross on her Crown and Cape, she ain't no fool. Come with me, and smash the Prime Directive, come and rise in love with the Number 13 and travel to other **Penta-Hexa** and **Rhombic Triacontahedral** worlds.

DEDICATION
to LILY MOSES
whose phine love fractalizes me,
making the inside the same as the outside.

(Mural by Jain, assisted by Lily Moses: 4"x6" Tantric Couple)

the
Phibonacci Sequence
is dedicated
to all Mathematical
PsychoNauts.

The Phibonacci Sequence is dedicated to all Mathematical PsychoNauts.

Future application of this 24 Repeating Pattern will help curb eco-side and corporate tyranny. We can no longer breed a society of eco-illiterate children.

This book is a dedication to the current and next generation of eco-illiterate children whom we have bred.

It is our responsibility to put in place the new soul-resonating education. We need to teach that whatever happens to the Earth, happens to us. We have to stop cutting down the trees and killing whales and all the other litanies of destruction, and be prepared for a profound revolution of consciousness, transformation and transcendence.

108 is the Pin Number for the UNIVERSE

PIN CODE OF THE UNIVERSE
11 – 4 – 10   JAIN 108 Mullumbimby

"The miracle is not to walk on water.
The miracle is to walk on the green earth
In the present moment".

- Thich Nhat Hanh.

Painting by my daughter **Aysha Sun**, 2009.
This was the first time the Phi Code appeared artistically to the world.
Oils on Canvas. Painted as a major assignment at her Lindesfarne school in Teranora, far north NSW.

The Phi Code is akin to a geometrical signal, one that conducts light, a series of switches, sophisticated and intelligent circuitry transporting energetic information to restore your DNA, our original primal huManWoman Design.

CHAPTER 1

why 108 ?

a detailed listing
of the occurrences
of
sri one hundred and
eight

Magic Sum = 108

(The Magic Sum of the 3 columns, 3 rows and 2 diagonals = 108)

by jain 108
mullumbimby, australia, 2008

includes sections on:

PART 1

- **108 in MATHEMATICS**

- **108 in GEOMETRY**

PART 2

- **108 in EASTERN RELIGIONS & TRADITIONS in the:**
  a) – HINDU
  b) – GAYATRI MANTRA
  c) – JAIN RELIGION
  d) – 108 in BUDDHIST
  e) – 108 in OTHER SECTS
  f) – 108 in MIDDLE EAST
  g) – 108 in CHRISTIAN

PART 3

- **108 in ASTROLOGY (Eastern & Western)**
  a) – Hindu
  b) – Western

PART 4

- **108 in MARTIAL ARTS**
- **108 in LITERATURE**
- **108 in OTHER FIELDS**
- **References & CONCLUSION**
- **THE DIVINE 108 LETTERBOX**

FORWARD:

NB: This following article on the various meanings attributed to the cross-culturally recognized and anointed number 108 is a sample of my 30 year collection of Number Harmonics, called:
JAIN'S DICTIONARY of NUMBERS
aka **HARMONIC STAIRWAY**.
eg: all this type of information just on 108 can be equally found on other important numbers like 12, 24, 32, 60, 90, 144, 216, 288, 432, 864, 1008 etc.

The manifestation of this booklet on Shri 108 is the beginning towards compiling this 6 foot tall visual dictionary of harmonics.

It is currently being compiled by the members of
JAIN F.R.E.E.D.O.M.S INC9882763 Non-Profit Research Organization and requires your donations and generosity to make this available to the world. No such comprehensive dictionary exists. If you try to google any number, say the number 108, you don't get this information as neatly packaged as it is here.

Also, since this body of work is part of a **Research Institute**, I am at liberty to reference and quote from many different sources, acknowledging any works or graphics or formulas from books or websites or lectures attended, without any worry about copyright infringement. It is more the concept of sharing knowledge. The research given does not necessarily mean that the following unverified facts are gospel, it is found and shared, and can be dealt with later. I have acknowledged all information where possible.

PART 1

108 in MATHEMATICS and GEOMETRY

108 in MATHEMATICS

- **108** (number)

- Cardinal: **One Hundred and Eight**

- Ordinal: **108**[th]

- Factorization.
  The **Divisors: 2, 3, 4, 6, 9, 12, 18, 27, 36, 54**

- Roman numeral: **CVIII**

- Binary: **1101100** (since $108 = 1 \times 2^6 + 1 \times 2^5 + 1 \times 2^3 + 1 \times 2^2$).

- Duodecimal (12-based) system, it would be written as **90.**

- Hexadecimal: **6C** (by definition: In mathematics and computer science, Hexadecimal (aka Base 16, or hexa, or hex) is a numerical system with a radix, or base of 16. It uses 16 distinct symbols, most often the symbols 0-9 to represent the values from zero to nine, and A, B, C, D, E, F to represent the values ten to fifteen).

- One Hundred and Eight is an **abundant number** (because the sum of its divisors listed above exceed the number 108) and a **semi-perfect number** (because some of these divisors add up to the sum of 108).

- It is a **Tetranacci Number**. (Just like the Fibonacci Number Sequence that starts with 1, 1 and keeps adding the previous number, the Tetranacci Sequence keeps adding the last 4 digits: 0, 1, 1, 2, 4, 8, 15, 29, 56, **108**, 208 etc.

- It is the **Hyper-factorial of 3** since it is of the form where $3^3 \times 2^2 \times 1^1 = 108$. ie: 27x4x1=108.
  The next Hyper-factorial Number of 4 is written like this:
  $4^4 3^3 2^2 1^1 = 27,648$.

- 108 is a number that is divisible by the value of its **φ function**, which is 36. (this means that according to **Euler's phi**

(φ) function or Totient, there are 36 numbers, from 1 to 108 that are prime to 108. This is called CoPrime eg: 6 and 35 are coPrime, but 6 and 27 are not coPrime because they are both divisible by 3. The number 1 is coPrime to every integer).

• 108 is also divisible by the total number of its divisors (12), hence it is a **Refactorable Number**.

(Regarding the symbol "Euler's phi (φ) function" it is mathematically confusing that Euler uses the Phi symbol that belongs to 1:1.618033... for something totally unrelated. It is better to leave Phi unadulterated, and let the mathematicians look for another symbol).

• In base 10, it is a **Harshad Number** (meaning that it is divisible by the sum of its digits, derived from the sanskrit word "harsa" meaning "great joy").

ie: 108 = 1+0+8 =9 which divides 108.

• 108 is a **Self Number**. (by definition, a Self Number or Colombian Number cannot be generated by any other digit added to the sum of its digits. eg: 21 is not a Self Number since it can be generated by the sum of 15, and the digits comprising 15,
ie: 21 = 15 + 1 + 5).

• In a **new western form of Gematria**, where a=1, b=2, c=3... j=10, k=11, l=12...x=24, y=25 and z=26 using the whole 26 letters of the alphabet, without compressing two digit numbers down to a single digit from 1 to 9, a number language is formed where every word we know, has a numerical value. Here, "**THE SPIRAL**" adds up to 108, and "**GEOMETRY**" adds up to 108.

• 108 equals the **sum of the first 9 multiples of 3**, viz. 0, 3, 6, 9, 12, 15, 18, 21 and 24.

• **108 Grains** = 36 Sicilicus (Roman Weights and Measures).

• $$3^3 + 4^3 + 5^3 = 6^3 = 6.6.6 = 216 = 2 \times 108$$

This is one of my favourites, as it takes the 2-Dimensional 3-4-5 Right Angled Triangle to the next dimensional level. This is pure number theory at its best. Its important too as the above cubic numbers appear to the student as shapes, that 3 different cubic forms when all put together rearrange themselves to make a larger cubic form.

(Bruce Cathie would say that this "double 108" relates to the physics of **Matter and Anti-Matter**, the Visible and Invisible, thus the Harmonic of 216 appears to be more wholistic towards understanding the nature of subtle energies).

- The **Magic Square of 3x3**, when its initial or starting number is 32, has a Magic Square Constant or **Magic Sum of 108** in its 3 columns, 3 rows, and 2 diagonals. Double the centre is 72, and if you examine the diametrically opposite pairs around the centre, they all add up to 72 eg: 33+39=72

37	*32*	*39*
38	36	34
33	40	35

Perhaps another alternative to this Magic Sum of 108, is to create a similar 3x3 grid and supply it with 9 numbers that have a total of 108. By selecting the consecutive or sequential or natural order counting numbers from 8 to 16, we create a Magic Square of 3x3 whose **Magic Sum is 36**, but the "Sigma" or sum of all numbers involved is 108. The name therefore for the following talisman I call "**Sigma 108**":

13	*8*	*15*
14	12	10
9	16	11

- 1080 divided by 666 = 1.621621621… or 1.<u>621</u> repeater. In Gematria, there is a lot of associated meanings to these numbers, particularly to the Moon (1080) and Sun (666) respectively.

- We learnt on page 38 of The Book Of Phi, Volume 3, that the final 2 digits in the Fibonacci Sequence has a periodicity of **300**! It is quite interesting that, since there is also a hidden 24 Repeating pattern in the digitally compressed Fibonacci Sequence, that the sigma or sum of the first 24 numbers also adds up to 300.
ie: 1+2+3+4+5+6+…+23+24 = 300
The shortcut or mathematical formula to determine this mentally is: $[n.(n+1)]$ divided by 2.
Thus for n=24, the answer is (24x25)/2 = 300.

|

OMNIA EX UNO
All Numbers Descend From One

108 in GEOMETRY

- Here is an unusual connection with 108 and 666. The Magic Square of 6x6, in 2-D has all its Magic Sums being 111 which means the sum of all digits from 1 to 6^2 is $1+2+3+...+35+36 = 666$. On a 3-Dimensional level, the Magic Cube of 6x6x6, whose Magic Sum is 651, has **108 columns** and 4 great diagonals.

Here are the 6 lots of 6x6 magic planes that make the Magic Cube of 6x6x6:

4	139	161	26	174	147
85	166	107	188	93	12
98	152	138	3	103	157
179	17	84	165	184	22
183	21	13	175	89	170
102	156	148	94	8	143

1

193	58	80	215	39	66
112	31	134	53	120	201
125	71	57	192	130	76
44	206	111	30	49	211
48	210	202	40	116	35
129	75	67	121	197	62

2

18	153	136	163	23	158
99	180	1	82	104	185
181	19	95	176	171	9
100	154	149	14	90	144
167	5	108	189	172	10
86	140	162	27	91	145

3

207	72	55	28	212	77
126	45	190	109	131	50
46	208	122	41	36	198
127	73	68	203	117	63
32	194	135	54	37	199
113	59	81	216	118	64

4

155	20	150	15	169	142
101	182	96	177	88	7
6	87	106	187	92	173
141	168	160	25	11	146
151	16	137	83	105	159
97	178	2	164	186	24

5

74	209	69	204	34	61
128	47	123	42	115	196
195	114	133	52	119	38
60	33	79	214	200	65
70	205	56	110	132	78
124	43	191	29	51	213

6

(taken from "Magic Square and Cubes" by W.S. Andrews , 1917, Penguin)

• Here is a geometric understanding of fractality, the ability to implode or make the Inside the same as the Outside. The diagram shows how the Golden Mean Spiral travels from the Macro Universe down to the Micro Atomic realms in a self-organized manner using 108°. Notice how the masculine straight lined spiral interacts with the feminine Equiangular Phi Spiral, where every angle hitting the spiral is the same **108°**, going from the long wave to the central short wave. This 108° angle permits suction to the centre, and infinite, non-destructive travel. It is the ultimate Compression Wave.

The 108° is found in the diagram where it says "theta".

(the diagram is sourced from the immense works of **Dan Winter** a Phi Fractal Scientist).

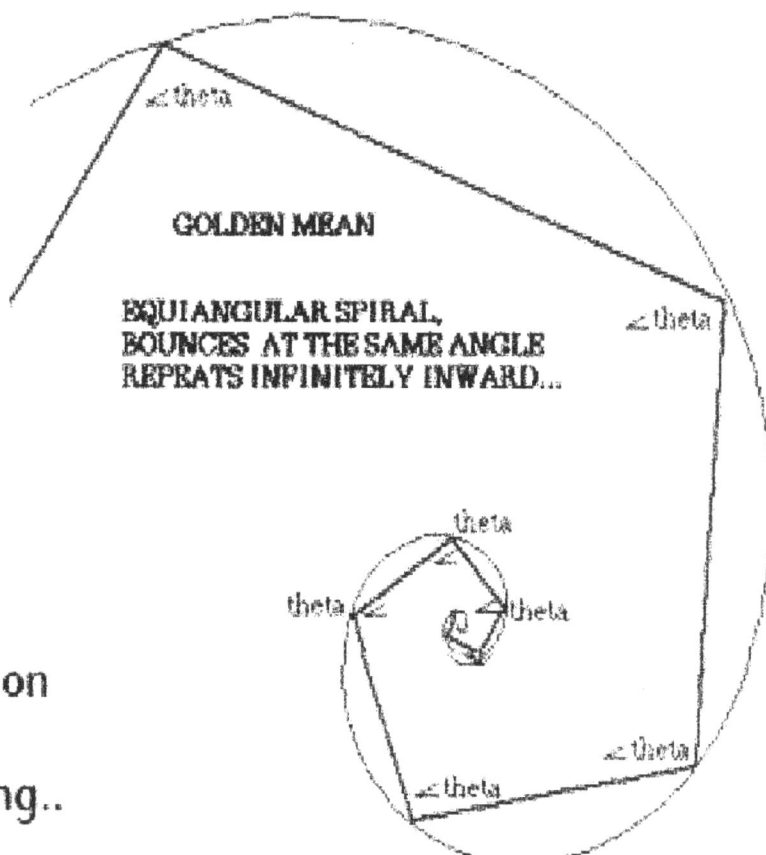

The ULTIMATE Compression Wave?.. for Charge, For Life Force.. For Matter Creation (out of charge) For Gravity Making..

GOLDEN MEAN

EQUIANGULAR SPIRAL, BOUNCES AT THE SAME ANGLE REPEATS INFINITELY INWARD...

A side Note: Having 360° is only an arbitrary number or choice. It could have been any number. It is not universal. The number 7 does not divide into 360 so what use is it. The first number that is divisible by all the numbers from 1 to 9 is **2520** (proposed either by Buckminster Fuller or Iona Miller of USA, you can google: The Auric Key or Syndex Synergetics). This makes more sense, as now we can see numbers as proportions, and not degrees.. So looking at the above diagram I am more interested in seeing what is the universal or forever proportion of this 108°. Let us divide 360 by 108 =180/54 =20/6 =10/3 =3.33333 Thus as a circular proportion, the 108° has a fixed ratio of 1:3.333...

In support of the 360 Circle Division, Bruce Cathie advocates the right-angled triangle having the sides: "**216, 288, 360**" which when halved create the Pyathagorean Triple:: "**108, 144, 180**". His emphasis is on the 144 Harmonic, Speed of Light. (see Page 31 of this book).

The ultimate expression of angles is not degrees but **Radians**. Forget the concept of 360° and start studying the ancient concept of Radians which is universal, in the sense that it takes the radius of 1 unit and traces this along the curve of the Unit Circle, and expresses everything now in a number proportion not in degrees! This is the key, viewing the open radius of 1 unit arching across a curve that is also 1 unit in length, see diagram:

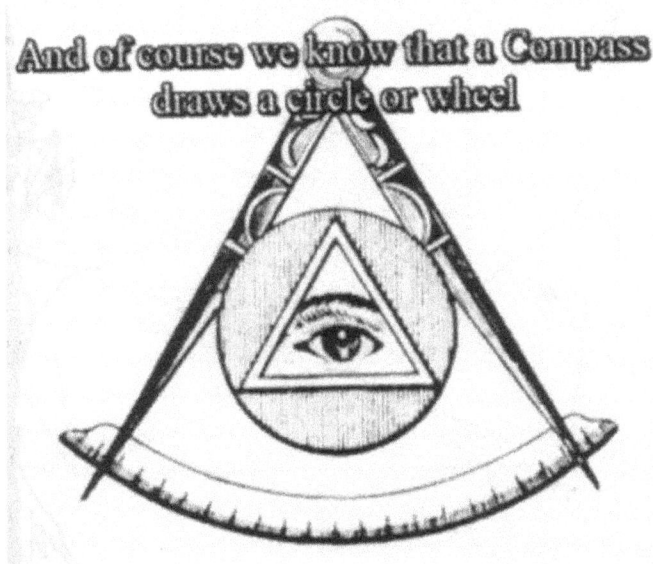

And of course we know that a Compass draws a circle or wheel

- There is an important 3-Dimensional shape called the **Lesser Maze**, from Theosophical Teachings, where the 5 Platonic Solids are embedded one within the other, naturally nesting according to Nature's rules. (This includes the Double or Star Tetrahedron, Cube, Octahedron, Icosahedron and Dodecahedron). There are **108 angles**, based on the apparent 36 equilateral triangles that have 60 degrees all around.
(The 36 is arrived by adding all the equilateral triangles: The Star Tetrahedron has 24 angles, The Octahedron has 24 angles and the Icosahedron has 60 angles; ie: 24+24+60=108). (The Cube has square faces, and the Dodecahedron has Pentagonal faces, so they are not counted in this series).

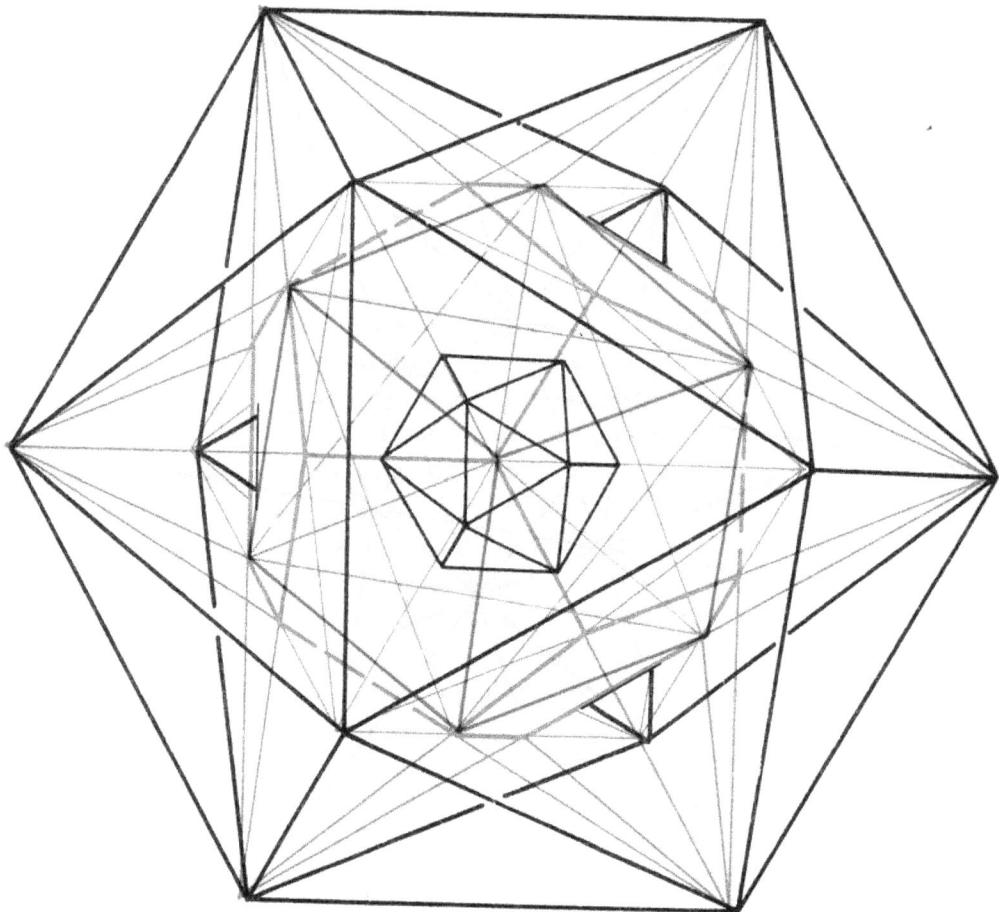

(Source taken from "The Mathematics Of The Cosmic Mind" by L. Gordon Plummer; 1970, Theosophical Publishing House).

- There are **108 free polyominoes of order 7**.

(By definition, a polyomino of Order 7 aka Heptomino or 7-ominos is a polygon in the plane of 7 equal-sized squares connected edge-to-edge. Here is a diagram of the 108 free **heptominoes**, "free" meaning that no rotations or reflections are included in this count.

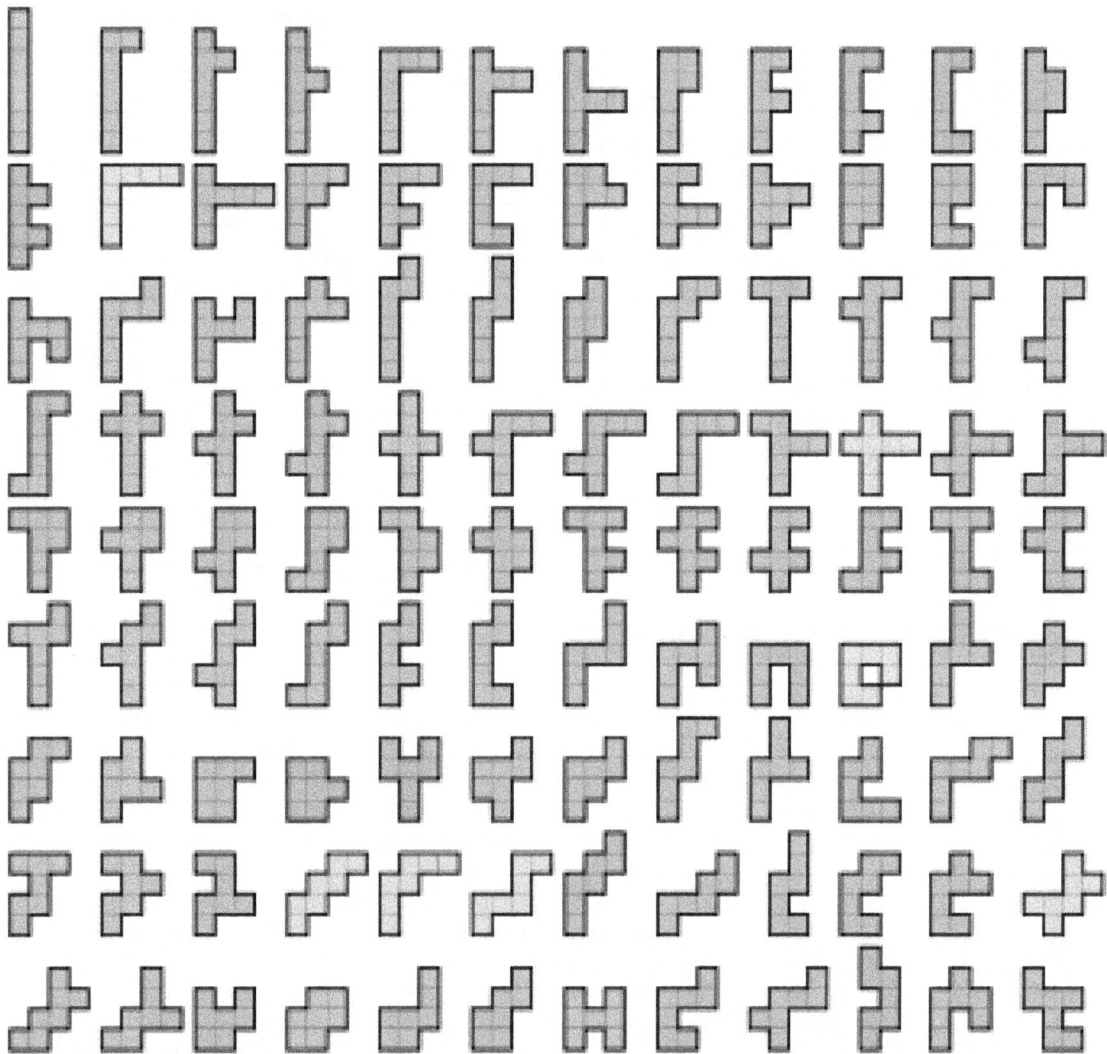

- There is an important **right-angled triangle**, whose base is 144 (The Grid Speed Harmonic of Light) can only be achieved or realized when the vertical height is 108 and the hypotenuse is 180. This can be proved by Pythagoras' Theorem that
$$108^2 + 144^2 = 180^2.$$
This is therefore known as a **Pythagorean Triple**:
108, 144, 180.
It helps us realize that if we double these numbers we get another set of triples: 216, 288, 360 that still obey the basic Pythagorean principle that $3^2 + 4^2 = 5^2$ another Right-angled triangle.

I believe that the two greatest mathematical gifts are:
1 – the 3-4-5 Triangle and
2 – The Golden Mean 1:1.618033... which has a lot to do with the 108 frequency being discussed here.

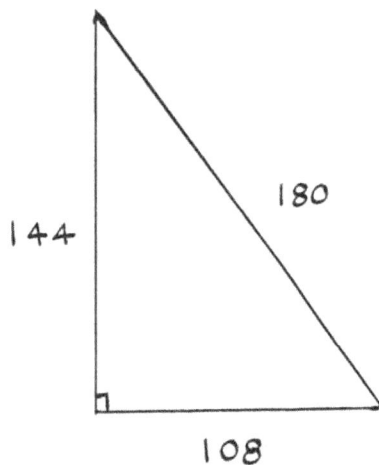

Higher Harmonic Octaves of the 3-4-5 Right-Angled Triangle:
$$108^2 + 144^2 = 180^2$$

(Source taken from Charles W. Leadbeater 33° book "The Hidden Life In Freemasonry" Theosophical Publishing House ,1926).

- Here is a **HyperDimensional Magic Cubic** with a twisted frame optical illusion yet all the numbers shown that range from 1 to 43, has **11 special rows** or columns that **all have the same sum of 108**.

eg: 8+24+29+36+11=108

It was composed by Arlin Anderson.

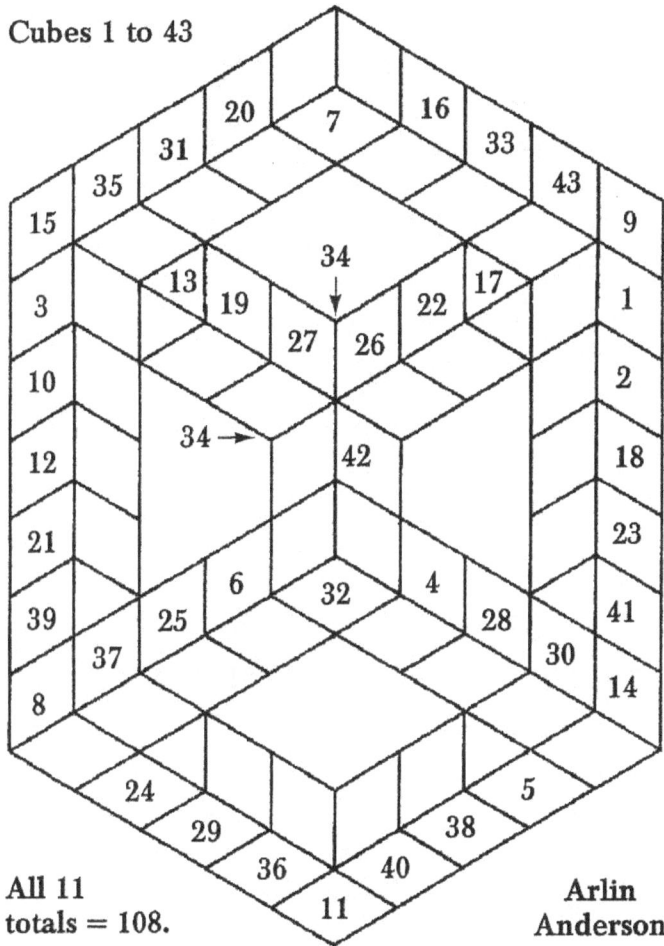

Cubes 1 to 43

All 11 totals = 108.

Arlin Anderson

(b)

Hyperdimensional twisted frame.
(b) Magic cubic treatment with sum 108.

• 2 MAGIC SPHERES WITH MAGIC SUMS OF 108

(taken from:p150 of Magic Squares and Cubes" by W.S Andrews, Dover, 1917)

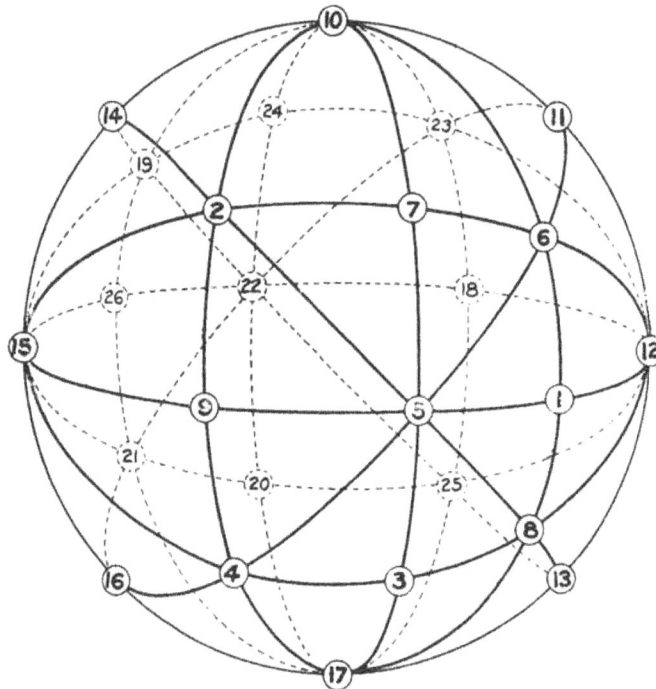

This remarkable Magic Sphere utilizing the consecutive numbers from 1 to 26 (like our Alphabet!) has **9 Great Circles** or Meridians running around the Globe and each Meridian is composed of **8 numbers**.

Lets examine one of these Meridians, the one sloping upwards at a 45° angle where

13 + 8 + 5 + 25 + 22 + 2 + 19 + 14 = **108**.

Observe all the Pairs of Numbers that are diametrically opposing one another, that is, if you passed a skewer through the globe passing through the centre the two numbers opposite add up to **27**.

(Is there a mathematical sequence that unites both 27 and 108? The only one that I know of is the Binary form of 27, where 27 is the initial number that becomes a Doubling Sequence as in:

27 – 54 – **108** – 216 – 432 – 864 – etc

or in other words 27 is a factor of 108 since 27 x 4 = 108).

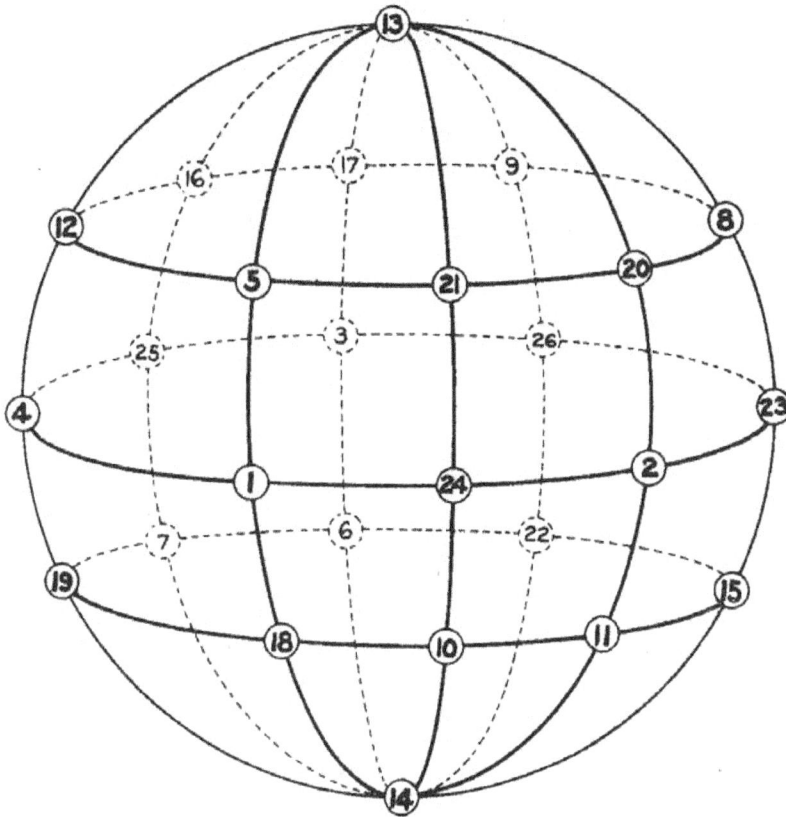

ANOTHER MAGIC SPHERE WITH MAGIC SUMS OF 108
(taken from:p150 of Magic Squares and Cubes" by W.S Andrews, Dover, 1917)

This remarkable Magic Sphere utilizing the consecutive numbers from 1 to 26 (like our Alphabet!) has **7 Great Circles** or Meridians running around the Globe and each Meridian is composed of **8 numbers**.

Lets examine one of these Meridians, the horizontal equator, where
1 + 24 + 2 + 23 + 26 + 3 + 25 + 4 = **108**.

Observe all the Pairs of Numbers that are diametrically opposing one another, that is, if you passed a skewer through the globe passing through the centre the two numbers opposite add up to **27**.

- In normal space, the interior angles of an equilateral **pentagon** measure **108°** each.

Shown here is not the pentagon form, but the pentacle or pentagram, sometimes known as the PentAlpha, showing its external angles of 108°.

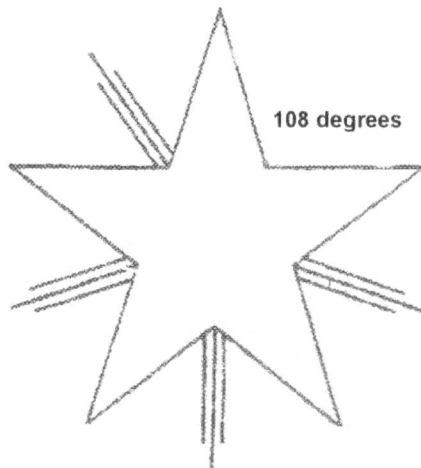

108 degrees

(As an aside, this is not an important observation, since this 108° is dependent on an arbitrary choice of dividing the circle into 360 divisions. It could have been divided into any number. What is therefore more important is to look at proportional relationships. eg: In the Golden Isosceles Triangle, we have two base angles of 72°. These angles written in degrees are not important, but its relationship of 360 divided by 72 = 5 is more important, thus that which is universal, the proportion of 1:5 has more significance).

If you chose to study further mathematics, examine the meaning of **"Radian"** which is far superior to measuring in man-made arbitrarily chosen angles, but is based on the universal forever radius of the circle, where if radius = 1 unit, and it is laid over the curving circumference of the circle, how much of an angle is made relating back to the centrepoint. 1 radian is about 57°. Mathematicians prefer this "radian" as it makes measurements and calculations much easier.

Aya of Sedona, has completed an incredible series of 108 large canvas paintings, 6' by 6' in size, known as:

The 108 STARWHEELS:

Here is an example:

"Turtle Isle", has a wheel of 24 motifs.

http://www.starwheels.com

The new Online School of Sacred Geometry.

www.schoolofsacredgeometry.org

PART 2

108 in EASTERN + WESTERN RELIGIONS

This includes:

a) – 108 in Hindu + Sri Yantra
b) – 108 in the Gayatri Mantra
c) – 108 in the Jain Religion
d) – 108 in the Buddhist
e) – 108 in Other Sects
f) – 108 in the Middle East
g) – 108 in Christian

"Sanat Kumara" by Lily Moses

a) - 108 in HINDU

- **JAPA MALA** or *Japa beads*, made from **Tulasi** wood. Comprising of 108 beads in total + the **head bead**. Sacred within Hinduism, Buddhism and connected yoga and dharma based practices.

- **Meru: This is a larger bead**, not part of the 108. It is not tied in the sequence of the other beads. It is **the guiding bead**, the Head Bead, the one that marks the beginning and end of the mala. (I believe that this Meru Bead, relates directly to the Double Nine Pair in the mathematical 108 Phi Code 1 Sequence shown below:

"PHI CODE 1" of 12 COMPLEMENTARY PAIRS OF 9											
1st Set of 12 Numbers	1	1	2	3	5	8	4	3	7	1	8	9
2nd Set of 12 Numbers	8	8	7	6	4	1	5	6	2	8	1	9

- **9 times 12**: Both of these numbers have been said to have spiritual significance in many traditions. Both Base 9 and Base 12 are galactic. 9 times 12 is 108. Also, 1 plus 8 equals 9. It is no wonder then that 108 (=9x12) is also **9 Dozen**!

- **Hindu deities** have **108 names**. A **Mala** usually has beads for **108 repetitions of a mantra**. Recital of these names, often accompanied by counting of 108-beaded Mala, is considered sacred and often done during religious ceremonies. The recital is called **NamaJapa**.

- Sanskrit alphabet: There are **54 letters in the Sanskrit alphabet**. Each has masculine and feminine, Shiva and Shakti. 54 times 2 is 108.

- According to the **Srimad Bhagavatam**, Krishna's celestial city has **108 petals**!

- The total of all digits of 108 (1+0+8) is **9**, which in Hinduism is said to represent the 9 tattvas. If you divide 108 by 2 or multiply by 2 the total of all digits again equals 9.

- It is described in the Srimad Bhagavatam that **Krishna dances with 108 'Gopis'** (cow-herd girls) in His Vrindavan pastimes, and later marries 16,108 wives in His city of Dwarka.

Hare Krishna devotees thus hold 108 as a number of great significance.

- **Siva Nataraja** dances his cosmic dance in **108 poses**.

- The small measurement of **one finger crease** (looking at your fingers there are 3 divisions or creases per finger,) which is about half an inch, when multiplied by 108 gives the precise height of each person, according to their specific finger crease size.

- Ananda Coomaraswamy holds that the numerology of the decimal numeric system was key to its inception. 108 is therefore founded in Dharmic metaphysical numerology. One for **bindu**; **zero** for **shunyata** and **eight** for **ananta**.

- The Lankavatara Sutra repeatedly refers to the **108 steps**.

- **Heart Chakra**: The chakras are the intersections of energy lines, and there are said to be a total of **108 energy lines** converging to form the heart chakra. One of them, sushumna leads to the crown chakra, and is said to be the path to Self-realization.

- **Marmas**: Marmas or Marmastanas are like energy intersections called chakras, except have fewer energy lines converging to form them. There are said to be **108 marmas** in the **subtle body.**

- **1, 0, and 8**: 1 stands for God or higher Truth, 0 stands for emptiness or completeness in spiritual practice, and 8 stands for infinity or eternity.

- Stages of the **Soul**: Said that **Atman**, the human soul or center goes through **108 stages** on the journey.

- **Breath**: Tantra estimates the average number of breaths per day at 21,600, of which 10,800 are solar energy, and 10,800 are lunar energy. Multiplying 108 by 100 is 10,800. Multiplying 2 x 10,800 equals 21,600.

- **108 Arhats** or Holy Ones.

- **108 Holy Places** for Vaishnavas, in Hinduism

- **108 Divyadeshes** - Divine or **Sacred Tirtha** throughout India and Nepal.

- **10800 bricks** in a **Vedic Hindu Fire Altar.**

- In the Indian epic **Ramayana** there are **108 offerings** that **Ram** was supposed to make. The earlier orally transmitted

Ramayana through disciplined oral transmission is 3,000 years old.

- A 108 story from the "**Tantra Shastra**" on the **108 Pitha** or **Sacred Places**:

The story goes that **Lord Shiva** was in deep and incessant meditation. His asceticism was creating great heat in the universe. All existence was in peril and Lord Brahma was deeply concerned. Lord Brahma asked the Mother of the Universe, Maa Shakti, to use Her strength and wile to seduce Lord Shiva. Maa Shakti agreed and was born as Sati, daughter of Shri Daksha. Lord Shiva was so entranced by Sati's asceticism and extraordinary beauty that he took human form and they were married. Years later, at a feast, Sati's father insulted Lord Shiva. Sati was so humiliated that she began a deep meditation which led to her immolation. Lord Shiva was completely heart broken. He reached into the sacrificial fire and pulled out as much of His beloved's body as he could grab. As He ascended to heaven, bits of Sati's body fell to earth. **108 bits of Sati** to be precise! In time, these places were acknowledged and worshipped.

- **108 Names of Lord Ganesha** and the meanings.

Here are the first 10 :

1. Akhurath: One who has mouse as his charioteer

2. Alampata: Ever eternal lord

3. Amit: Incomparable lord

4. Anantachidrupamayam: Infinite and consciousness personified

5. Avaneesh: Lord of the whole world

6. Avighna: Remover of obstacles

7. Balaganapati: Beloved and lovable child

8. Bhalchandra: Moon-crested lord

9. Bheema: Huge and Gigantic

10. Bhupati: Lord of the gods

- **Goddess names**: There are said to be 108 Indian goddess names.

- **108 Upanishads** as listed in the Muktikopanishad.

There are **4 Vedas** or categories:

Rigveda (has 10): eg: Aitareya , Atmabodha, Nirvana etc

Yajurveda (has 50): eg: Amrtanada, Brahmavidya, Yogakundalini

Pancabrahma, Satyayani, Hamsa, Turiyatita etc.

SamaVeda (has 16): eg: Samnyasa, Vasudevai etc.

Atharvaveda (has 32): eg: Atma, Krishna, Garuda, Ganapati, Sarabha, Annapurna, Tripuratapani, Devi, Bhavana, Sita, etc.

- The **Rig-Vedic verses** are conventionally counted as having approximately **10800** stanzas.
The Rig Veda has 10,800 stanzas with 40 syllables per stanza, a total of 432,000 syllables. There are **10,800 bricks in the Indian fire altar** (Agnacayana), a funeral pyre, a number of fate.

- Many Hindu monks carry titles like:
"Swami **1008** Padmananda".

- There is an extraordinary temple at Konark, Orissa, the **Lingaraja Temple** a **complex of 108 temples**, and a Gold and White Buddha Pagoda called the Dhavaigiri Stupa.

- The **Elkinji** temple also has **108 temples** within its walls.

- **Kshatriya Dhangars** have **108 clans**. The lineage of these clans are based from the solar and lunar dynasties.

- 108 Percussions required when making Eastern and Western Homeopathics; this system was originally developed in ancient India. This means the medicine bottle is tapped on the palm of the hand 108 times.

- 108 therefore signifies the **wholeness** of divinity or perfect totality.

- **Sahasra-nama-stotra** or Sahasranama (1008 names derived from a list of 10,000 names that were given to a rishi called Tandi by Shiva; the rishi gives it to Upamanyu and the latter gives it to Krishna in an abridged form. He gives only one tenth of the names which form the 1008 names) has a total of **108 shlokas**.

- In the **Ramayana, the monkey Hanuman** breaks the skull of the demon Lighting Tongue into **108 pieces**.

- **Sri Yantra**: On the Sri Yantra there are marmas where three lines intersect, and there are 54 such intersections. Each intersection has masculine and feminine, Shiva and Shakti qualities. 54 x 2 equals 108. Thus, there are 108 points that define the Sri Yantra as well as the human body.

In the world of Sacred Geometry, it is useful to visualize these 9 intersecting triangles as 9 inter-penetrating **pyramids**.

Stan Tenen of www.meru.org has made a beautiful wire frame of this. The diagram below is a shadow of a multi-dimensional form.

According to **Randy Masters** of California, the Sri Yantra or Sri Chakra, is the **spectral emission of the Hydrogen Atom**. Every line we see in this yantra is actually one of the Powers of Phi (which as I have shown is called Phi Code 2 with an infinitely recursive patterning based on 108).

(Artwork of Sri Yantra computerised by Wolfram Oehler of Germany, who gifted me his artwork, for staying in my studio, temple of mathematics).

- **108 OMS** on the 108 Petals of this flowery mandala that has a 3-Dimensional Flower-Of-Life pattern in the main large circle:

108 in EASTERN RELIGIONS & TRADITIONS
b) - 108 in the GAYATRI MANTRA

• The most famous 108 repetition is of the **prayer for enlightenment** called the **Gayatri Mantra**. It is meant to be the oldest known prayer, and has a numerological value in Sanskrit of 108, this number being based on the 24 syllables of the Gayatri Prayer. Now, what is it that relates to both 24 and 108? That's right, the "Phi Code 1" table shown above.

It is therefore important to learn the words to this Prayer:

> **"Om Bhur Bhuvah Swaha / Tat Savitur Varenyam,**
> **Bhargo Devasya Dhimahi / Dhiyo Yo Nah Pracho-dayat"**

which has many translations:
"This prayer appeals to the highest Wisdom, to the Brilliance of the Cosmos, to illumine our understanding as individuals and also as a World Family. The appeal requests that we become subtle and receptive to the Divine Wisdom which pervades the experience of consciousness..."
By chanting this 108 times (or smaller increments of this based on 9 or 18 or 27 or 54) one is purified.

• An important part of the Phi Code Mysteries is its sonic encodement into a Universal Prayer known in the Vedas as the Gayatri Mantra which is always chanted 108 times, (and sometimes 54 or 18 times which are factors of 108). It has 24 syllables. Advisable to chant it at dawn, noon and dusk, so long as a total of 108 have been completed. Your friends and family and wealth can pass, are illusionary, but Gayatri is eternal.
It is a complete mantra for adoring God or seen as an address to the energy of the sun, an Inner Sun and an Outer Sun. Gayatri is all Gods in One, a Universal Prayer.
Just like the Phi Code, which is a trinity, the Gayatri, as the Mother of the Vedas, consists of 3 feminine deities: 1- Gayatri is the master of the senses, 2- Savitri is the master of the life force and signifies Truth, and 3- Saraswati is the master of speech. It is therefore a blessing of Unity and Purity of Thoughts and designed to help spiritual aspirants on their path to perfection.
It was a way of hiding this ancient knowledge, and storing its deep

structure memory into a song, so that the masses would sing it, be filled with bliss, but not fully understand its cryptic meaning.

I met the head of an orange-robed RamaKrishna order, I am currently banned from actually stating who he is, but when we had a luncheon together in Kuala Lumpur where I happened to be running seminars, I asked him one burning question, having informed him of my mathematical derivation of Shri 108. I asked him "Why 108 in the Gayatri Mantra? and he replied that it is based on 24 syllables and has a numerological sum of 108. Now, what in the universe has a basis of 24 and 108. There is only one thing I know, with 2 variations, and that is the mathematical compression of the linear Fibonacci Sequence and the mathematical compression of the multi-dimensional Powers of Phi!

So lets look at this Gayatri Mantra and several translations of this song.

OM BHUR BHUVAH SVAHA
TAT SAVITUR VARENYAM
BHARGO DEVASYA DHEEMAHI
DHEEYO YO NAH PRACHODAYAT

OM = Para Brahman
BHUR = Physical plane, earth
BHUVAH = Middle world, atmosphere
SVAHA = Heaven
TAT = Paramatma, God
SAVITUR = Source – that from which all this is born
VARENYAM = Fit to be worshipped
BHARGO = The radiance, the spiritual Effulgence
DEVASYA = Divine Reality
DHEEMAHI = We meditate
DHEEYO = Buddhi, intellect
YO = Which
NAH = Our
PRACHODAYAT = Enlighten

Translation 1: (from Sai Baba Order)
We meditate on the spiritual effulgence of that adorable supreme divine reality, the source of the three worlds – physical, subtle and

causal, May that supreme Divine Being stimulate our intelligence so that we may realize the supreme truth.

Translation 2: (from Swami Sukhabodhananda)
You live in a Universe and there is a Universe within you.
There is an external sun and an internal sun. Gayatri Mantra blesses us with Divine Energy from both of them.

Translation 3:
Let us contemplate that admirable radiance of the God Savitr. May he direct our minds.

(In this 3rd translation, the Hymn is to a god (a vivifying god of the sun, pictured as a chariot driver, however the hymn itself is considered a goddess- Gayatri Devi. Later, Gayatri was changed to Kali!).

Many devotees believe that this mantra will also cure our diseases. It also acts as a Third Eye to reveal to us the Inner Vision of Bhagwan or God. It can be chanted mentally or verbally, like breathing air day and night.

If a Theosophist, like Charles Leadbeater was to translate this hymn, they would invariably see it as an Invocation to the Sun or Solar Logos, where shafts of the rising or setting sun fill the reciter with the 7 Rays, therefore acting as a prism and extending the aurae of the chanters.

श्री गायत्री यंत्र

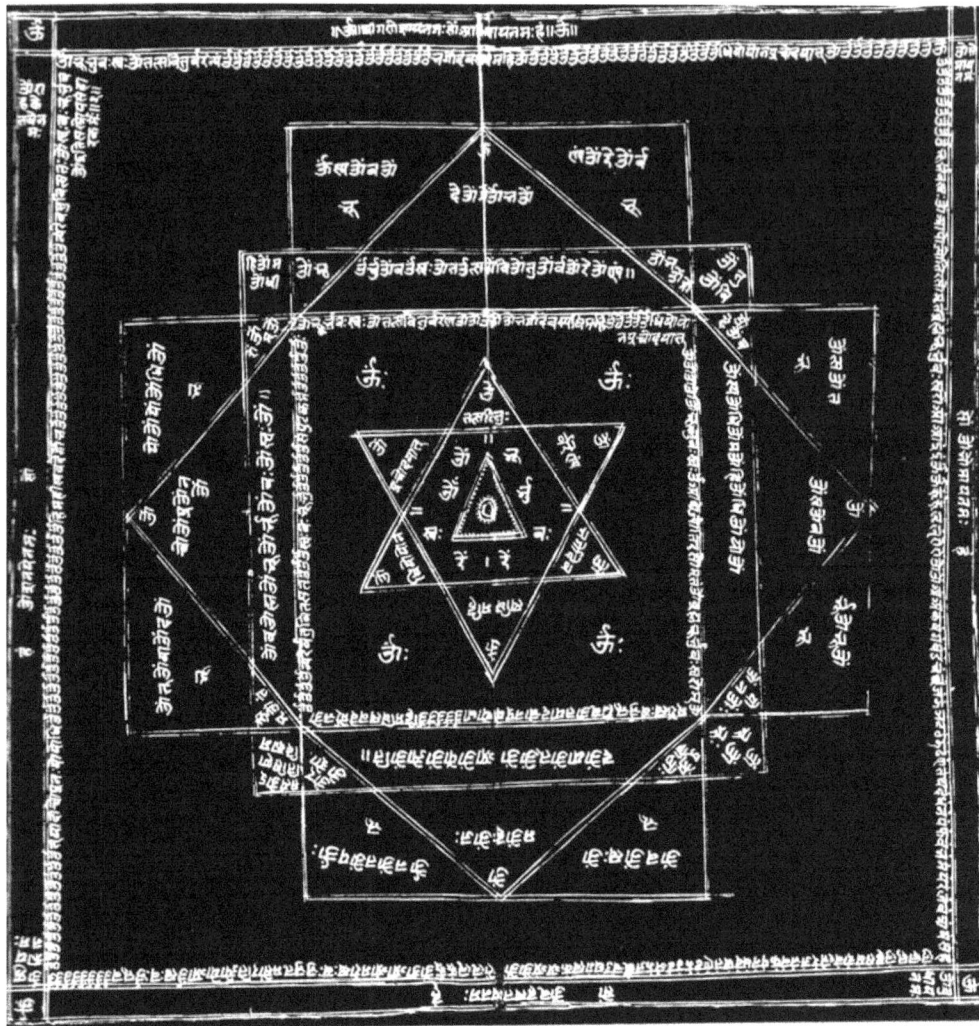

The Gāyatrī Yantra inscribed with the Gāyatrī Mantra, Rajasthan, c. 19th century. Ink and colour on paper

**The Gayatri Yantra inscribed within the Gayatri Mantra:
Rajasthan, circa C19th, ink and colour on paper
Notice that at the centre is the Phi ratioed Triangle!**
(reprinted from "Yantra" book by Madha Khanna, pub. 1979)

108 In OTHER EASTERN TRADITIONS & RELIGIONS

c) - 108 In the JAIN RELIGION

● **Jaina Time Cycles of 24**: The following is a reference to 24, not 108, but it is interesting that the Jain religion highlights both these numbers: 24 and 108. There were 24 Jain Teachers, none living at the moment. Here is an image of the Jaina Time Cycles:

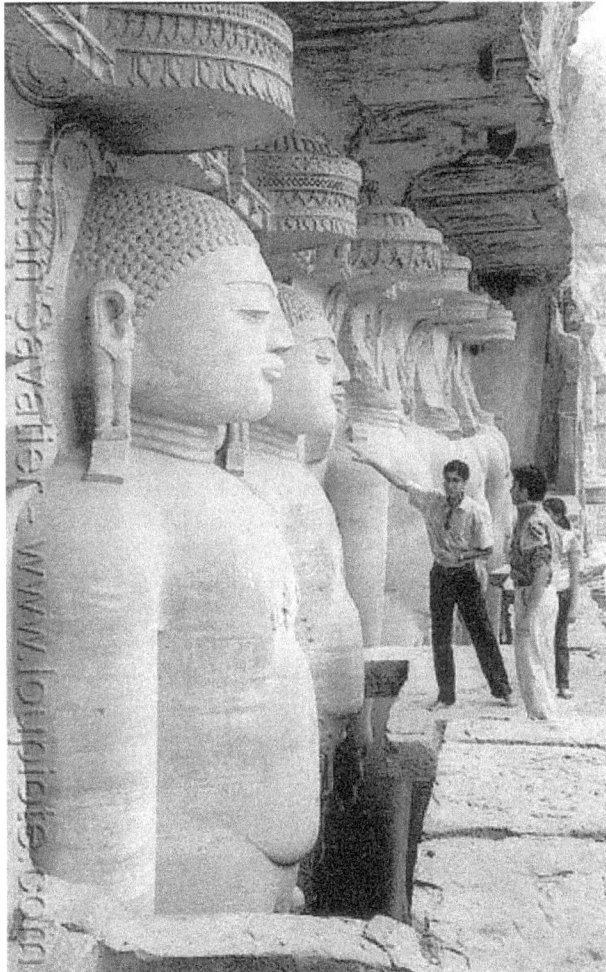

The 24 Jain Teachers, carved in stone in Gwalior.

24
PAST
24
3
6
9
12
15
18
22
24
PRESENT
24
FUTURE
TIME
CYCLE
धर्म
जिन
जैन
जीव
त्रिकाल चौबीसी
(सम्पूर्ण जैनत्व)

Jain Time Cycles as Wheels of 24,
hinting at the Phi Code of 24.

Jain Emblem of Hand, showing 24 rays issuing from the Palm Centre. The Indian flag also has 24 Rays in a circular motif.

- In the **Jain religion**, 108 are the combined virtues of five categories of holy ones, including 12, 8, 36, 25, and 27 virtues respectively. (12+8+36+25+27=108).

- The number of sins in Tibetan Buddhism.

- Many Buddhist temples have **108 steps**.

- All Buddhists accept the **Buddha Footprint** with its 108 Auspicious Illustrations. These areas are considered to have been marked on the Buddha's left foot when his body was discovered.

- The **Chinese Buddhists** and **Taoists** use a 108 bead mala, which is called "**Su-Chu**", and has three dividing beads, so the mala is divided into three parts of 36 each.

- There are various coloured types of rosary beads for helping the Tibetan Monk to count his 108 prayers. Yellow string or beads is used to invoke Buddha; white string or white beads made from sea-shells is for Bodhisattva; red strings or coral beads is for the one who converted to Tibetan Buddhism.

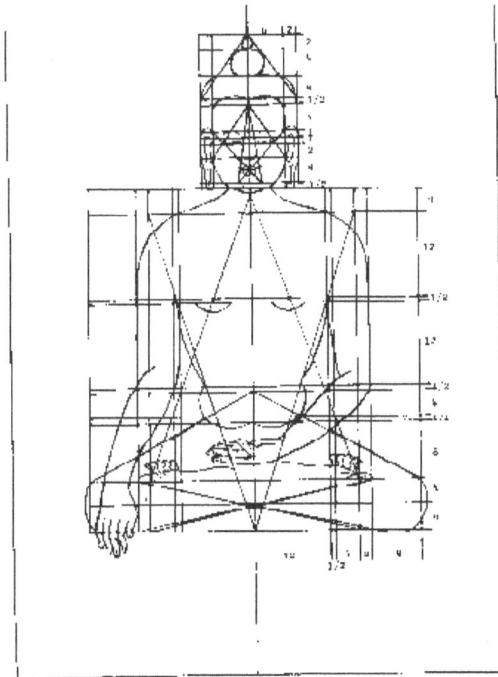

e) 108 in OTHER SECTS

- **108 Sacred Water Taps** in Muktinath-Nepal

- **108 Seats** of the **Nepalese Parliament.**

- In the temple **Angkor Wat** area, Cambodia, there are numerous references to the number 108, which plays a significant role in the symbolism of the structure. **Angkor Wat incorporates the numbers 54, 108 and 540, in its planning, architecture and engineering**. Angkor was laid out for both practical transportation and for honor to celestial deities. The city has a diameter of about two miles, and was surrounded by a moat with five bridges. It has five gates, and to each of them leads an avenue, bridging over that water ditch which surrounds the whole civic area on a flat plane. **A row of huge stone figures, 108 per avenue, 54 on each side, a total of 540 statues of the Indo-Aryan deities Deva and Asura, border each of these roads, and each row carries a huge Naga serpent with nine heads.** (Patten and Spedicato).

- In **Japan**, at the end of the year, a bell is chimed 108 times to finish the old year and welcome the new one. Each ring represents one of 108 earthly temptations a person must overcome to achieve nirvana. **Omisoka** is the day of **New Year's Eve** in **Japan**, one of the biggest celebrations in the year where people eat Toshikoshi-soba at night and stay up till midnight to listen to the 108 chimes of a nearby temple bell.

- **Zen priests** wear juzu (a ring of prayer beads) around their wrists, which consists of **108 beads.**

- **The Tale of Genji** is a great classic of Japanese literature, written by Murasaki Shikibu (c. 980- c.1014), a complex story of life and loves of the shining prince Genji, developed in some thousand pages **divided into 54 books**.

- **Bowing exactly 108 times** is a tradition **in Japan**...the obeisance of one hundred and eight bows.

- The **Sikh tradition** has a mala of **108 knots** tied in a string of wool, rather than beads.

- The birthday of the **Prophet Noble Drew Ali** is on **January the 8th, written as 108**. African Moors in the northwest have Ethiopian ancestors going back for thousands of years. The prophet Ali, in the early C20th helped awaken the former slaves to uphold their dignity from being classified as negro, black, coloured or Ethiopian.
(email information supplied to me by Raymond Bey, a Moorish American, 2003).

- There are **540** (ie: 108 x 5 = 540) **doors to the Great Hall of Valhalla** in Nordic mythology.

- In **Uxmal**, in the **Yucatan Peninsula**, the "**Pyramid of the Magician**" has two tall, wide staircases each having **54 steps.**

- In the ancient architectural planning of the city of **Machu Picchu**, high in the Andes, in Peru, (time of construction not known with certainty), there were twelve quarters and 216 buildings.
On this ancient **calendar** there appear **54 jaguar heads** and **108 condors**.

- **Golok woman** from Changthang **(the Tibetan plateau),** have waist-length **hair braided into 108 plaits** to reflect the 108 blessings of Buddha. (Patten and Spedicato)

- **In China...**The boat he took on the Yangtze Kiang had **a dormitory with 108 beds...** (Patten and Spedicato)

- In some parts of **Central Asia** ... deeply influenced Tibetan Buddhism, Clark relates how he visited **a Ngolok chieftain, in his tent, where 108 lamps were burning** in front of the statue of a Bon divinity. (Patten and Spedicato)

- Camels in caravans bringing tea from **China to Siberia via Mongolia** were loaded with two or three boxes of tea bricks; **each box contained exactly 108 (Chinese) pounds of tea.** (Prjevalski)

- Traditional gambling in **Tibet** used not dice, but **108 small carved bones in a hollowed out skull.**

- The main temple of **Lhasa** is the Zuglacan; **here 108 episodes of the life of Buddha** are represented by frescoes.
(Tucci).

- One of the oldest temples in **Tibet,** founded in the 8th century by **Padmasambhava,** is located in Samye. (or was, when Tucci visited it half a century ago; we don't know if it is one of the few Tibetan temples that survived destruction during the Cultural Revolution). **The temple contained 108 chapels;** not far from it another temple, the Ngari Tratsang, contained frescoes of the **108 "works"** of Buddha. (Tucci).

• **Muktinath** is a temple town **in Nepal,** where Buddhism and Hinduism coexist with several temples. Near the Vishnu temple the water of a holy source is distributed by **a system of 108 outlets.** (Tucci).

(Stephano Rotondo, C12th Italy)

f) 108 in MIDDLE EAST

- **Islam:** The number 108 is used in Islam to refer to God.

- From a recently translated tablet, dealing with creation myths (Pettinato, communication at Accademia dei Lincei meeting, Rome, June 2000) we know **that Enki gave 108 "essences" to Inanna.** Now essences is a term referring to spiritual capacities that finally end up in man. **This may be the earliest reference in literature to the number 108.** (Patten and Spedicato)

- **Gudea, a Sumerian king** of the city of Lagash, built a temple to Ningirsu **employing 216,000 workers (see Sitchin [4], p.45 in Italian edition by Edizioni Mediterranee, 1996). His meal usually consisted in 108 different servings**.

- In offerings to the gods **in Uruk 108 types of common dates** were included.

- **The Book of Enoch**, a canonical biblical book for the Christians and the Hebrews of Ethiopia and Armenia, maybe also for the Essenes in Qumram, but not included in the present Christian or Masoretic canon, **consists of 108 chapters.** In 1947 **Athanasius Yeshue Samuel,** Metropolitan of the Syrian Orthodox Archidiocese in Jerusalem, bought from bedouins the first four scrolls of the rich cache from Qumram. It included a complete book of Isaiah, **written in 54 columns,** of 30 lines each. See Samuel [30].

- The **Da'wa system** of **28 Arabic Letters** with corresponding numerical values and associated divine attributes with names, meanings and numerical values for the translation of the word values. In this table below, **the value of 108** is associated to the letter "Ha" the 8th letter of the Arabic alphabet, and has the attribute of "**HAQ**" which means "**Truth**"; thus Truth relates to the vibration of 108, in this Arabic system.
An Arabic expression spoken by Jesus is: "**Unna el Haq**" meaning "I Am the Truth".

LETTERS		VALUES	ASSOCIATED DIVINE ATTRIBUTES			VALUES
			NAMES		MEANING	
ا	'alif	1	الله	ALLAH	Allah	66
ب	ba	2	باقي	BÁQÍ	He who remains	113
ج	jim	3	جامع	JÁMI'	He who collects	114
د	dal	4	دیان	DAYÁN	Judge	65
ه	ha	5	هادي	HÁDÍ	Guide	20
و	wa	6	ولي	WALÍ	Master	46
ز	zay	7	زكي	ZAKÍ	Purifier	37
ح	ḥa	8	حق	ḤAQ	Truth	108	←
ط	ṭa	9	طاهر	ṬÁHIR	Saint	215
ي	ya	10	یسین	YASSÍN	Chief	130
ك	kaf	20	كافي	KÁFÍ	Sufficient	111
ل	lam	30	لطیف	LAṬÍF	Benevolent	129
م	mim	40	ملك	MALIK	King	90
ن	nùn	50	نور	NÙR	Light	256
س	sin	60	سمیع	SAMÍ'	Listener	180
ع	'ayin	70	علي	'ALÍ	Raised up	110
ف	fa	80	فتاح	FATÁḤ	Who opens	489
ص	ṣad	90	صمد	SAMAD	Eternal	134
ق	qaf	100	قادر	QÁDIR	Powerful	305
ر	ra	200	رب	RAB	Lord	202
ش	shin	300	شفیع	SHAFÍ'	Who accepts	460
ت	ta	400	توب	TAWAB	Who restores to the good	408
ث	tha	500	ثابت	THÁBIT	Stable	903
خ	kha	600	خالق	KHÁLIQ	Creator	731
ذ	dhal	700	ذاکر	DHÁKIR	Who remembers	921
ض	ḍad	800	ضار	ḌÁR	Chastiser	1,001
ظ	ḍha	900	ظاهر	DHÁHIR	Apparent	1,106
غ	gha	1,000	غفور	GHAFÙR	Indulgent	1,285

FIG. 20.51. *The Da'wa system, after the tabulation made by Sheikh Abu'l Muwwayid of Gujarat in Jawahiru'l Khamsah*

(This table of the Da'wa system is made after the tabulation by Sheikh Abu'l Muwayid of Gujarat in Jawahiru'l Khamsah).

- **Baalbek (Lord Baal),** is the name of a very ancient city **in Lebanon.** Its Greek name was **Heliopolis.** It is located on the way from Beirut to Damascus, inland from the Mediterranean Sea. Formerly it was a city of great size, and its ruins cover a large area. The ruins consist of buildings of several periods, Roman columns and structures having been built over much more ancient structures, including huge megalithic blocks, among the biggest known in the world. Long predating Roman and Greek structures on the site, the three that make up the so-called Trilithon are as tall as five-story buildings and weigh over 600 tons each. A fourth megalith, abandoned in its quarry before completion of the cutting, is almost 80 feet in length and weighs 1100 tons. Amazingly these giant blocks were cut, perfectly shaped and somehow transported to Baalbek from a quarry several miles away. In addition they were skillfully incorporated, at considerable height. **The temple had 54 massive columns during Phoenician times.** (Patten and Spedicato)

"**Baalbek**" offers the puzzling question: How did the ancient Phoenician builders raise this lintel, this giant horizontal stone upon the original 9 columns, is the largest single uncut stone in the world, and there is no machinery today that can lift such a weight!

g) 108 in CHRISTIAN

Shown here is the **Chalice Lid of the Glastonbury Well**, depicting the well-known Vesica Piscis of two interlacing Circles. There is a vertical line that runs through the centre of the 2 circles. This line can represent the Number "**1**". The Outer Circle encompassing the whole geometry represents the number Zero "**0**", and the two interlacing circles represents the number "**8**". Put all together as one number code, it reads as **108**. This is pure speculation on my part (Jain), but I insert this information here just for the record.

- Perhaps more surprisingly number 108 seems to have made some inroads even in modern **catholic world**. For instance a **Virgin Mary golden statue** has been put on the top of Milan cathedral in Italy at the height of **exactly 108 meters**. One of the encyclicas written by the present **Pope Paul John II contains 108 chapters**. One of his ordainings of bishops consisted of **108 individuals**.....

- **Zosimus** was a pagan officer **in Rome** early in the C6th A.D., about whom little is known, except that he was clearly against the Christian emperors who had introduced new religious rites, while the empire was under great crisis. He was an admirer of emperor Julian. In his work "New History", in 13 chapters, there is yet another indication of an **ancient reverence for the numbers 54**

and 108. He referred to secular games that were held in Rome on occasions somewhere between 105 and 110 years apart.

The location was the Field of Mars in the northern part of the city of Rome. **The games were organized following a ritual described in the Sybillian books**, albeit the text was somewhat corrupt and not easily understood at his time. Among the elements of the ritual, hymns and peans were sung by 27 young ladies and 27 young lads, who were chosen under the condition of being amphithaleis, i.e. both their parents had to be alive. **Thus the ceremony involved 54 youngsters and 108 parents.** (See Zosimus) and (Patten and Spedicato)

- In Celtic Britannia, there are over 300 henges in England, Scotland and Ireland. Of particular interest is the **henge near Ruthven, in Scotland**.

Inside the big ring of large stones is a second circular ring of smaller "kerbstones". Our count from a map of this site is **108 kerbstones**, set very close together, but this needs local site confirmation. (Patten and Spedicato)

- D. Patten when referring to an educational movie about Stonehenge, states that **the Celtic architects used a 54-year period** for multiple year unit, as we use "decade" or "century". In the movie the narrator stated that "nobody knows why" such a 54-year period was used. (Patten and Spedicato).

PHI
CODE
108

PART 3

108 in ASTROLOGY: EASTERN & WESTERN

108 in EASTERN or HINDU ASTROLOGY

- In Western Astrology, the 360° of the circle is subdivided into an equal house system of 30 minutes each sign, thus
360 x 30 = **10,800** minutes of arc.
In the Vedas, Zero {0} is considered 'Purna' or complete. So we take out the last zeros and are left with 108. The idea of our total universe is represented by this number of 108. Offering 108, devotees believe that they are showing ultimate or complete respect to the Supreme.

- **Vedic Astrology:** There are 12 constellations, and 9 arc segments called Namshas or ChandraKalas. 9 times 12 equals 108. Chandra is moon, and kalas are the divisions within a whole.

Planets and Houses: In astrology, there are **12 houses** and **9 planets. 12 times 9 equals 108**.

- Hindu Kshatriya Dhangars have 108 clans. The lineage of these clans is from solar and lunar dynasties.

- It should be noted that the diameter of **the Sun is 108 times the diameter of the Earth**. The distance from the Sun to the Earth is 108 times the diameter of the Sun.
The average distance of the Moon from the Earth is 108 times the diameter of the Moon.

- **River Ganga**: The sacred River Ganga spans a longitude of 12 degrees (79 to 91), and a latitude of 9 degrees (22 to 31). 12 times 9 equals 108.

- Ancient Hindu astrology places the **limit of human life** at **108 years**

- The Hindu number: 432,000, the number of years attributed to a **Yuga**, or a world age, also happens to be equal to the number of guardians of the **Germanic Walhalla** or heaven (viz. 800 for every one of its 540 gates), can be analysed as **108** x 4000.

- Early Buddhism and Hinduism on the basis of the study of the **eclipses** referred to as **saros cycles** proposed the cosmology of cycles of worldly creation and destruction and postulated for every 108 years. (The Assyrians and Babylonians called it the "**Sharu Cycle**" of **18 years** and 11 and 1/3 days, and Edmund Halley named it the saros cycle. Current astronomy has settled for average **18.2-year cycles**). We are interested in this stellar fact since: 6 x 18 = 108

- 108 in the "**Sanatana Dharma**". In a book by Khurana, the original **Vedic Harmonics of Time** appear:
A circle has 360 degrees, which when multiplied by 60 gives us **21,600** minutes in a circle. 60 comes from the **60 'ghatis'** which Sanatana Dharmiks believe in. **One ghati is equal to 24 minutes** and 60 ghatis come to 24 hours. **One ghati is divided into 60 parts or 'palas'.**
So the 60 ghatis multiplied by 60 palasa comes to **3,600**.
This is further multiplied by 60 (because **a pala contains 60 vipalas**) which gives us 21,600.
Half of this is for the day, and the other half for the night. So, 21,600 divided by 2 gives us **10,800**. For practical purposes, we use 108. Using the number 108 helps us coordinate the rhythm of time and space & we remain in harmony with the spiritual powers of Nature.

- The Indian celibate sect, the **Brahma Kumaris** have a rather cryptic and **cosmic sequence:**

8   **108**   16   **108**   900,00   33,000,000.

108 in WESTERN ASTROLOGY

- **Silver and the moon**: In astrology, the metal silver is said to represent the Moon. The **Atomic Weight of Silver is 108.** And listed as the **Atomic Number** of **Hassium.**

- According to a Russian lecturer I met at a Nexus Conference in Brisbane in 2005, Professor **Valery Uvarov** claims there is a second planet similar to Earth, behind the Sun. He talked of the Heartbeat of our Planet, and that the Breath of Earth is 54 minutes In, and 54 minutes Out, giving a total of **108 minutes for a one cycle Breath of Earth**. This was true, according to ancient texts, when Earth originally had an orbit of 360 days. Memories of this knowledge are currently stored in the Pyramid of Gizeh!

- There is a **Mayan Time Code** or astrological cycle of Venus over 8 years that refers to the **108 passes of Venus over our Sun**, which is termed one cycle. The next cycle ends in 2012.

- According to researchers Patten and Spedicato, the planet **Mars passed close to Earth every 54 years** during the Catastrophic Era which terminated in the year 701 BC. This was first proposed in the controversial work "Worlds in Collision", published by **Velikovsky** in 1950. Based on many RetroCalculations, there arose a hypothesis that catastrophes, of extraterrestrial origin, occurred at periodical intervals of 54 and 108 years.

PART 4

108 in MARTIAL ARTS

- Marma Adi has **108 pressure points**.

- The Chinese school of martial arts agrees with the South Indian school of martial arts on the principle of 108 pressure points.

- 108 number also figures prominently in the symbolism associated with Karate, particularly the Goju Ryu discipline. The ultimate Goju-ryu kata, Suparinpei, literally translates to 108. **Suparinpei** is the Chinese pronunciation of the number 108, while gojushi of gojushiho is the Japanese pronunciation of the number 54. The other Goju-ryu kata, Sanseru (meaning "36") and Seipai ("18") are factors of the number 108.

- Several different Taijiquan long forms consist of 108 moves.

- Paek Pal Ki Hyung, the 7th form taught in the art of Kuk Sool Won, translates literally to "108 technique" form. It is also frequently referred to as the "**eliminate 108 torments**" form. Each motion corresponds with one of the 108 Buddhist torments.

- The litmus test of the Indian presence is the set of number concepts that have been used in the architecture of **Chinese martial arts 3, 9, 12, 18, 36** and finally the umbrella of them all: **108**. Praying thrice to the Buddha, the 18 hands of lohan, the 36 targets of tien hsueh, the 108 of the Yang long form and Yip Man's compression of the Buddhist Ng Mui's gift into **108 movements** for the form are all signs of the intellectual legacy (not necessarily techniques) of Indian Buddhism.

- In Indian martial arts such as the South Indian Kalaripayit there are **108 strikes** to various **nerve centres**.

- The 108 of the Yang long form and Wing Chun, taught by Yip Man having **108 movements** are noted in this regard.

("108 Mandalog" by Iona Miller of USA,
proposing another dimensional view of Number Order)

108 in LITERATURE

• In **Homer's Odyssey**, 108 is the number of suitors coveting Penelope, wife of Odysseus.

• There are **108 outlaws** in the Chinese classic Water Margin/Outlaws of the Marsh.

• 108 is the number of Surat al-Kawthar in the **Qur'an**, the smallest one of the Book. (There are 114 Suras or chapters in the Koran). Al-Kawthar translates as Abundance, Plenty.

• There are **108 love sonnets** in **Astrophil and Stella**, the first English sonnet sequence by Sir Philip Sidney.

108 in OTHER FIELDS

- 108 is the **emergency number** in **India**.

- A Hare Krishna, or 'krishnacore' hardcore **punk band 108**.

- **108 acres** of the **Vatican City**, Italy.

- Western esotericists refer to 108 as the Beastly Sum of the squares $6^2 + 6^2 + 6^2 = 108$ the Biblical-Apocalyptic number of the Beast: 666

- Traditionally the **Angelus Bell** is **rung 108 times**.

- **Ben Turpin's** schtick for taking a fall (similar to Chevy Chase's) was known as a "**one-oh-eight**" or a "hundred and eight".

- Chinese astrology and Tao philosophy holds that there are **108 sacred stars**.

- **108 Stars** is also the name of a group of space pirates who use Tao magic in the anime Outlaw Star.

- The **number of stitches** on an American **baseball**.

- The **number of minutes** that Yuri Gagarin, a Soviet cosmonaut, orbited the earth during the **first manned space flight** on April 12, 1961.

- 108 is the ten code for officer down or **officer in danger**.

- The number of "missing" episodes of the BBC television program Doctor Who, destroyed in the 1970s.

- Squaresoft's Final Fantasy Tactics has an accessory called **108 Gems**, which cancels various status effects and enhances elemental magic spells.

- The number of **Mbit/s** of a non-standard extension of **IEEE 802.11g wireless network** using channel bonding.

- 108 is the name of a community of and for **open source developers**, created by Red Hat.

- The **Pokémon Spiritomb** seems to be intricately linked to the number 108.

- It is part of the Mythology in the video-game **The Legend of Dragoon**.

- Konami's Suikoden series always has 108 characters to recruit, called "**Stars of Destiny**".

- "**108**" is the name of an **Italian artist** born in 1978 famous for his street art and graffiti from Alessandria (90 miles from Turin). His first works known by people were enigmatic blob-like yellow shapes designed with a strong intention to make visual chaos.

- The **108 Hospital** is the name of the hospital in **Hanoi**, Vietnam.

- **Oscar Ichazo**, the student of Grudjieff had a **vision of 108 Ennegrams** in 1954 and subsequently set up a school in Chile, South America, in 1971, calling his system of archetypal divination "Arica".

- In Tokko (manga) 108 demons had to be killed in order to obtain 108 fragments necessary to be used to close the demon's portal.

- There are **108 cards** in a common deck of playing cards called **Uno.**

References

From **Wikipedia**, the free encyclopedia
http://en.wikipedia.org/wiki/108_(number):

"The **Penguin Dictionary of Curious and Interesting Numbers**"
by Wells, D.   London: Penguin Group. (1987): 134

External links

- The Significance of the number 108
- Meaning of 108 beads on a mala
- Article on the symbolism of the number 108

Retrieved from http://en.wikipedia.org/wiki/108_%28number%29

CONCLUSION:

I would like to dispute or bring to your attention, several writings on 108 where the authors actually state that 108 is not a magic number of the Universe (because they do not know the mathematical origins of Sri 108 as explained in these Phi Books), so I quote the following from one dismal source so that you can dismiss the nonsense:

[Much in the same way, the number '108' is just a reference frame. It is symbolic of a bigger picture: that of humility. When devotees recite 108 Hanuman Chalisas, in their minds they believe, they are proving their love for God, and that there is in fact a need to prove their love. When devotees assign 108 names to Shri Ganesh, they are once again gauging their devotion through numbers. This, of course, may be considered unreasonable, since it suggests that 108 chants are more effective than 109 chants. How do they know this? **Have they proved it? Is 108 the magic number of the universe? No, it is not! It is a reference frame.** What is important is that a system is imposed to guide us through the fundamental struggles encountered in any evolutional process. Otherwise, chaos and anarchy follow and nothing gets done].

A skeptical account: (taken from:
http://www.hknet.org.nz/108meaning.html

"The Significance of the number 108").

My response:

I have highlighted in bold above this skeptics most interesting comment: **Is 108 the magic number of the universe? No, it is not!**

If only he knew that there is indeed a great mathematical depth to the origin of 108. Its a good lesson in not trusting internet or On-line information from ignorant sources.

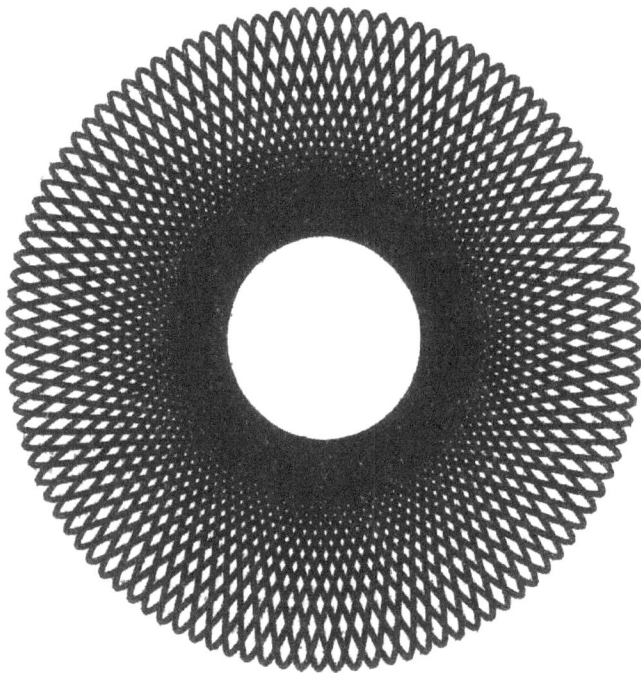

This Spirograph Wheel has 108 Points

THE DIVINE 108 LETTERBOX

I took these photos of someone's letterbox approaching Bondi Beach from the northern side, and was inspired by the vines growing around the brickwork.

It is quite rich in symbolism, vines and letterboxes. A "letter box", is simply our vessel that receives communication from the outside world into our inner world. It's a portal, a gateway.

And what of vines! I understand that the origin of the word "DIVINE" is "de Vina" meaning "of the Vine" which is really the perfect image of fractality, the ability to share branching wavelengths that do not cancel one another out.

Here is the long view, show the North Bondi home and its viney letterbox.

(nb: the digital time on the photo is 5:55, which is the number-plate of my camry car at the time).

Here is a close up view of "de vina 108"

These books on shri 108 have shown a clear connection between this frequency of 108 and its relationship to the Divine Proportion. It is therefore very apt to have the vine interacting with the 108 letterbox. There's a deep communication going on. The bricks are symbols of structure and building form. The Living Vine is communicating, is announcing the arrival of 108's revelations to the world, a discourse on the Living Mathematics of Nature.

ABOUT JAIN:

Jain's central problem or "spiritual core issue of concern" is that he is a man of the 22nd Century living in the 21st!

Poetical and Passionate, he is attempting to bring in the feminized mathematics based on 108 and the True Value of Pi, but is challenged appropriately and vigorously by rigid curriculum-makers who fear change and a loss of job. He is an intrepid **Mathematical PSYCHONAUT** and **mathematical FUTURIST**.

CHAPTER 2

108 PHI CODE 1
as 1 x 24 & 2 x 12 & 3 x 8 & 4 x 6
ARRAYS or MATRICES
REVEALING a HIDDEN 666 PATTERN

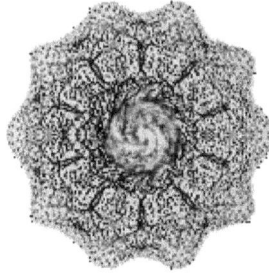

INCLUDES:

- **PART 1:**
  - — **The Infinite "108 -9" Phi Code Sequence as a 1x24 Array & as an 2x12 ARRAY or MATRIX**

- **PART 2:**
  - — **PHI 108 Code Alternate Number Sequence Arranged in a 4x6 Frame**
  - — **Practical Application of the 108 Phi Codes**

- **PART 3:**
  - — **The Compressed 24 Repeating Pattern Expressed as Trinitized Octaves or Rectangular Arrays of 3x8**
  - — **The 216 Code**

Shown here is 1mx1m oil painting by Aysha Jain Sun, submitted as her Year 12 project, 2009. Notice the Wheel of 24 Compressed Single Digits of the Fibonacci Sequence.

This Nature woman is thus the Embodiment of the Phi Code, aptly located around her Crown Chakra.
She holds an egg, symbol of creation, and the sunflower extends its roots into her subconscious. The emanating rings of circles, seen only as subliminal curves, are distanced in the Phi Ratio.

— The Infinite "108 -9" Phi Code Sequence as a 1x24 Array & as a 2x12 ARRAY or MATRIX

Based on Digital Compression or the Simplification or Distillation or Reduction to Single Digits of the Fibonacci Numbers:

1 – 1 – 2 – 3 – 5 – 8 – 13 – 21 – 34 – 55 – 89 – 144 – 233 - etc

Here is the Phi Code or 24 Repeating Pattern for the Reduced Fibonacci Sequence:
1, 1, 2, 3, 5, 8, 4, 3, 7, 1, 8, 9, 8, 8, 7, 6, 4, 1, 5, 6, 2, 8, 1, 9

Jain's 108 Phi Code: an Infinitely Repeating 24 Pattern Based on the Compression of the Fibonacci Numbers into Single Digits

1	1	2	3	5	8	4	3	7	1	8	9	8	8	7	6	4	1	5	6	2	8	1	9

Fig 1
THE 1 X 24 ARRAY

It was demonstrated in previous articles that the above 24 Repeating Pattern can be viewed as two halves:

.........PHI CODE of 12 COMPLEMENTARY PAIRS OF 9.........											
1st Set of 12 Numbers	1	1	2	3	5	8	4	3	7	1	8	9
2nd Set of 12 Numbers	8	8	7	6	4	1	5	6	2	8	1	9

Fig 2
This is the 2x12 ARRAY

Notice that overall, there are 12 Pairs of 9 which = 108, except for the last pair that is a double 9 or a double Pair, which seems to act as a bridge or a bond for the infinitely continuous 108 code.
It appears, that the ancient secret goes like this:
108 – 9 – 108 – 9 – 108 – 9 or tabulated as:

Infinite "108 -9" Phi Code Sequence								
108	9	108	9	108	9	108	9	108

Fig 3

The 108 Code or Necklace is really punctuated with Joins or Links
or Bridges that represent the "9" bead,
as in DNA that has a stop and start button that knows when to end
and start another chain of molecules.

Thus this is part of DNA's Numerical Architecture, so PHIne,
it is irreducibly part of our Astral Hygiene
or better expressed as HIGH GENE
(credit to my friend the Alchemist Teacher Michael Lamb
who told me about "High Gene". We're always looking for creative new word plays).

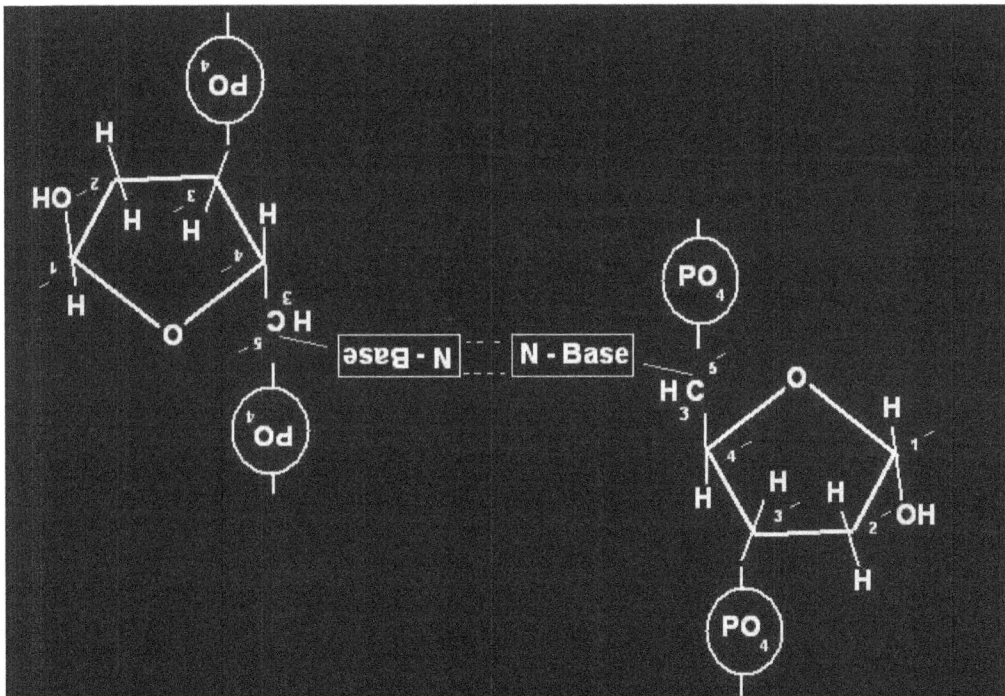

Figure: It is the actual structure of in which both strands are in opposite directions.
See C5' of right hand structure is upward and C5' of left side structure is
downward and C3' of right is downward and C3' of left is upward.
Also see that left side structure is not only inverted but also left to right
rotated to have H-bonding with opposite nucleotide. (Fig by Junaid Ahmad)

PART 2:
The 4X6 ARRAY

Underlined are the Alternative Numbers as shown in Fig 1

$\underline{1}$, 1, $\underline{2}$, 3, $\underline{5}$, 8, $\underline{4}$, 3, $\underline{7}$, 1, $\underline{8}$, 9, $\underline{8}$, 8, $\underline{7}$, 6, $\underline{4}$, 1, $\underline{5}$, 6, $\underline{2}$, 8, $\underline{1}$, 9

ALTERNATE NUMBERS IN THE 24 REPEATING PATTERN																							
1	1	2	3	5	8	4	3	7	1	8	9	8	8	7	6	4	1	5	6	2	8	1	9

Fig 4

These 24 numbers in Fig 4 will be plugged into a 4x6 frame. The first 2 rows will contain the above 12 underlined alternate numbers, and the following 2 rows will be filled in with the 12 remaining numbers.

PHI 108 CODE ALTERNATE NUMBER SEQUENCE ARRANGED IN A 4X6 FRAME					
1	2	5	4	7	8
8	7	4	5	2	1
1	3	8	3	1	**9**
8	6	1	6	8	**9**

Fig 5

Observe that the first 5 columns each have a sum of 18 (or 2 x 9), except for the last or 6th column which has a surprising sum of 27 (or 3 x 9).
It is possible to rearrange or rewrite the above sums of 18 as 6+6+6
And the last column of 27 as 9+9+9
This data can be plugged into a 3x6 Frame:

THE HIDDEN 666 MYSTERY in the PHI CODE 108 PATTERN					

6	6	6	6	6	**9**
+	+	+	+	+	**+**
6	6	6	6	6	**9**
+	+	+	+	+	**+**
6	6	6	6	6	**9**

Fig 6
The Secret Number 666
is found hiding here in the Living Mathematics of Nature.
It was embedded in the Alternate Sequence
or Every Second Number of the 24 Pattern, but revealed
only by the Magic Tool of **Digital Compression**.

Fig 6 shows another way of writing out the Phi Code, instead of summarizing it as an infinite repeating sequence of
108 / 9 / 108 / 9 etc.

It can seen that its Magic Sum of 24 repeating numbers is indeed the sum of 5 x (6+6+6) + 3 x 9 = **117**.

It suggests that if we strip away this Vedic fascination for the vibration of 108, that the real Magic Sum or Pulse of the Universes is 117, an interesting **Prime Number** that needs further research. In mathematical language, we say that the total sum, or Sigma ("**Σ**" one of the 24 Greek Letters) of the 24 Repeating Phi Pattern is:
Σ = 117

PRACTICAL APPLICATION OF THE 108 PHI CODES

These sequences can also be creatively applied to musical scales and colour frequencies etc to assist in the healing of the body, mind and spirit, and to raise the consciousness of the Planet.

Here are some various activities and crafts that can be designed to embed this knowledge into the creative arts, and ultimately into our Hearts.

1 – Make a Fashion Belt:
I would suggest to have 24 small Phi Ratioed Rectangles that are to be coloured in from a palette of **9 favourite colours**, so that the strict structure of the above sequences are obeyed by ascribing say 1 = Red, 2 = Orange, 3 = Yellow, 4 =Green, 5 = Blue, 6 = Indigo, 7 = Violet, 8 = Pink, 9 = Rose

2 – Make a Fashion Necklace or a monk's Mala Beads for chanting.
Again, specify 9 colours for the single 9 digits in the 24 repeating code, if the size of the beads are all the same size and shape.
If you like, you can use **9 different shapes** of beads to translate this code into a physical form.

3 – Make a Floor Tiling based on the 4x6 Frames shown above in Fig 5. Again, you can use **9 colours**, but since this is an experiment in consciousness, where there are no rules, and all possibilities, you could use one colour throughout but **having 9 different textures**.

4 – Make Phi Code Music based on the selection of 9 different and favourite notes that can be selected from a child's Xylophone. Instead of 9 individual notes which would give a linear or monotonous feeling to the music, you could select say 9 distinct chords.
You could look at the 12 Pairs of 9 and use this pattern, as shown in Fig 2, to combine 2 chords at a time to complete one cycle of the 2x12 frame.
Or, regarding the 4x6 frames, you could take 4 notes or 4 chords at a time using 4 people or musicians in the band to translate the numbers of Fig 5 into music.

5 – Do you have your own original suggestions that can translate the 108 Phi Codes into this earthly realm? literally to bring down Heaven to Earth, if you believe in duality, but if you understand Unity Consciousness, that is, how to **UNI-PHI**, there is no above or below, no heaven or earth, only Holo-Luminous interconnectedness and fractality as we begin this process of **Fibonaccization of the Human Species**, toroidally embedding this self-similar, infinitely recursive Mathematics of the Soul, into our Heart, our EarthHeart beat of 108, 108 108 Hertz, cycles per second.

Blessings from **Jain 108**

17-Nov 2007
(Bundjalung Beach, far north NSW, Australia)

"omnia apud me mathematica fiunt:

With me everything turns into mathematics."
Descartes

PART 3:
THE COMPRESSED 24 REPEATING PATTERN EXPRESSED AS TRINITIZED OCTAVES OR RECTANGULAR ARRAYS OF 3X8

As a creative procedure, it is possible to seek more patterns in the Phi Code, by looking at the factors of 24.

The factors of 24 are 2x12, 3x8, 4x6

Which indicates that we can split these 24 repeating numbers into a 3x8 rectangle or a 4x6.

This conjures up the image of Octaves or the **Law of Octaves**.

(We have already seen the Phi Code split into 2x12 rectangle or frame and also the 4x6).

The following information is just for the record, just to say that we looked at this 3x8 frame, curious as to what we might find as an rhythmophile (lover of rhythms or patterns). Some patterns were found and I leave it to the reader to go deeper down this rabbit hole, as truly there is much more beauty hidden in this 3x8 arrangement than meets the eye.

You will remember in Fig 2 that the original Phi Code 1 that has the 24 Repeating Pattern of:

1	1	2	3	5	8	4	3	7	1	8	9
8	8	7	6	4	1	5	6	2	8	1	9

was expressed as 12 Pairs of 9, or 2 Levels of 12 with a double bond on the last Pair where both numbers were 9,

thus the sequence really can viewed as:

(12 Pairs of 9) _9_ (12 Pairs of 9) _9_ (12 Pairs of 9) _9_ etc or written like this:

108 _9_ 108 _9_ 108 _9_ 108 _9_ 108 _9_ 108 _9 etc

Here is the Rectangular Array of 3x8:

1	1	2	3	5	8	4	3
7	1	8	9	8	8	7	6
4	1	5	6	2	8	1	9

Fig 7
3x8 Rectangular Array of Phi Code 1
showing repetition in the 2nd and 6th columns
of 111 and 888, highlighted.

My friend and Phi Teacher extraordinaire Aya of Arizona, has found a hidden and dazzling 666 pattern in this Figure 7 above, but for reasons regarding the amount of hidden symmetry in this matrix, a lot of this information is being withheld until I release my next and more advanced booklet just on this topic.

Just to give you a little bit more, it is best to understand that the Shri 108 Frequency is only half of this Unified Equation and the real secrets lie in seeing Fig 7 mirror-imaged, and to do this we must double this 108 code and review Fig 7 not as a 3x8 array but as a 3x16 rectangular array:

1	1	2	3	5	8	4	3	Mirror	3	4	8	5	3	2	1	1
7	1	8	9	8	8	7	6	Image	6	7	8	8	9	8	1	7
4	1	5	6	2	8	1	9	Array	9	1	8	2	6	5	1	4

Fig 8
3x8 Rectangular Array of Phi Code 1
Mirror-Imaged Symmetries
defining this now as a 2x108 or 216 Code

Fig 8 is relevant as it is akin to the Mirror-Imaging of DNA genetic information that utilizes Mirror-Imaged Sequences.

This therefore is where the real realizations are found. I did express this Secret of 216 as a conclusion in my first book "The Book Of Phi", Vol 1, self-published in 2002, page 145, that "**The Number is 216**".

Here is therefore indisputable evidence that the more we probe into the Fibonacci Sequence and the Powers Of Phi, we keep revealing innate, inherent, underlying order as the matrix of all creation.

In my next publications on the Phi Mysteries, I will start including some advanced computer renditions of charts and more organized data for your interest and reviewing.

Without going into too much depth, Fig 9 shows an example of the upcoming data to be supplied. This diagram was created by my dear and ENGRAILED Friend, another Fibonatic, Adrian Asfar of Victoria, Australia, who has contributed is genius into offering this upgraded data:

"The three-fold number is present in all things whatsoever,
nor did we ourselves discover this number,
but rather nature teaches it to us"...
- Ovid

1	2	3	4	5	6	7	8	9	10	11	12	13	14	15	16	17	18	19	20	21	22	23	24
1	1	2	3	5	8	4	3	7	1	8	9	8	8	7	6	4	1	5	6	2	8	1	9	1	1	2	3	5	8	4	3	7	1	8	9
-8	0	1	1	2	3	-4	-1	4	-6	7	1	-1	0	-1	-1	-2	-3	4	1	-4	6	-7	8	-8	0	1	1	2	3	-4	-1	4	-6	7	1

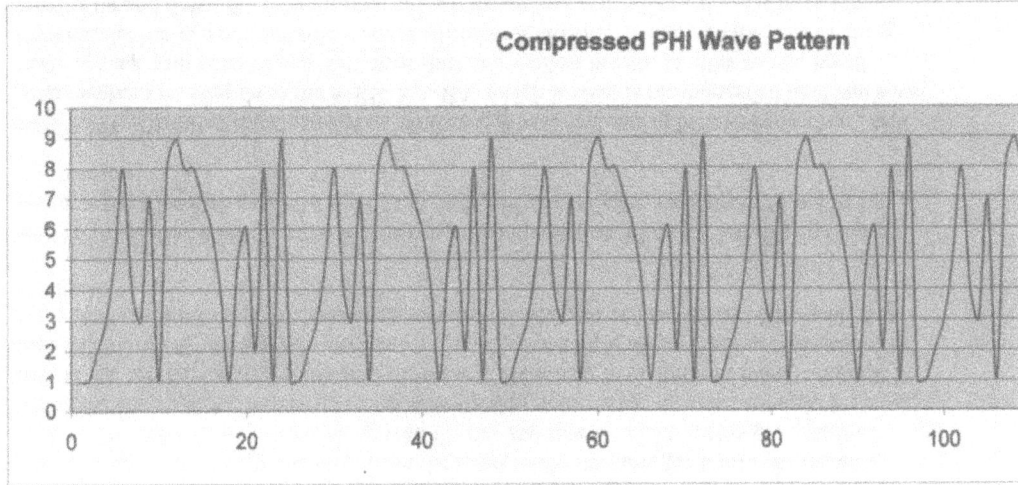

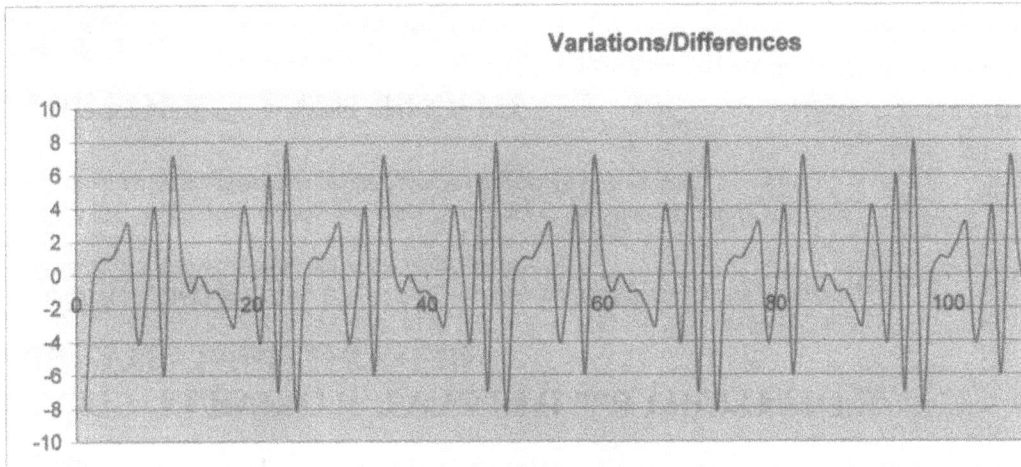

Compressed PHI Wave Pattern

Variations/Differences

Fig 9
Phi Code 1 plotted as Phi Wave Patterns
in many ongoing sets of 24,
to show the bigger recursive picture.

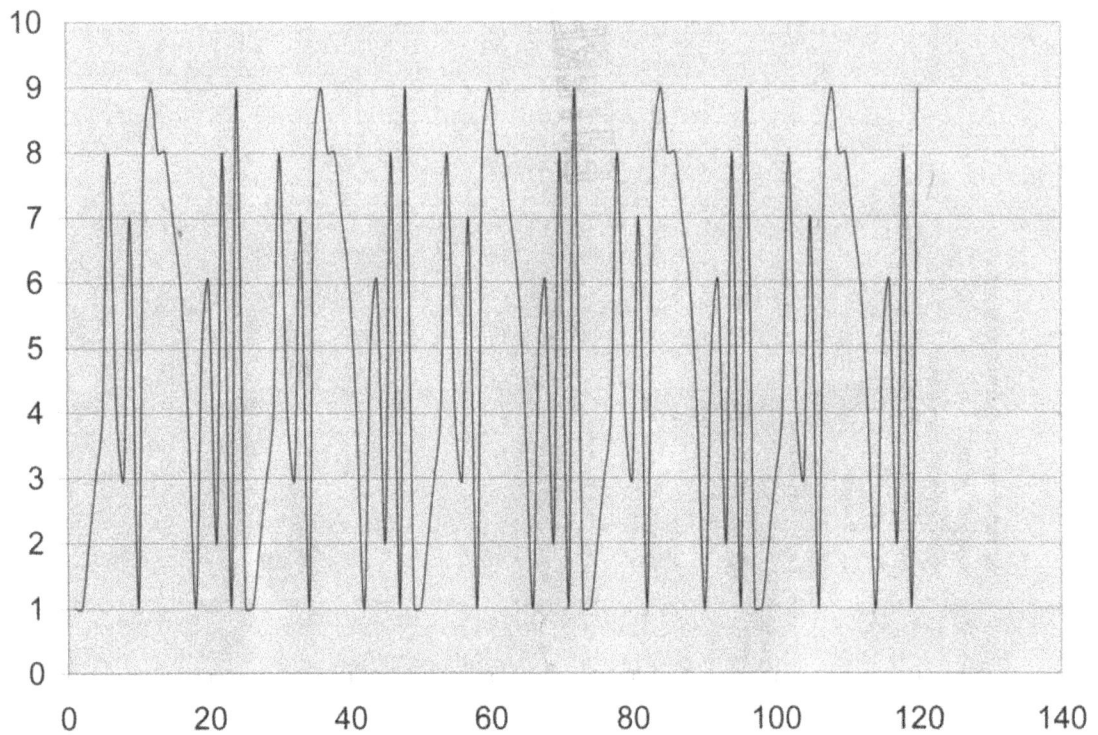

Fig 10
Phi Code 1
graphed up to the 120th Fibonacci Numbers,
to illustrate the obvious repetition every 24 digits.

(acknowledgment to Adrian Asfar for these graphs).

CHAPTER 3

JAIN'S PHI-PRIME CONNECTION

**The
PRIME NUMBER CROSS,
24-ness and
the PHI CODE**

by JAIN 108
2008, Mullumbimby Creek, Oz

- **PART 1:**
  — **Definition** of **Primes**
  — **Ulam's Rose**
  — The **TRANSLATION** Of **NUMBER** Into **ART**
    aka **The ART of NUMBER**

- **PART 2:**
  — **The Prime Number Cross** or the
  4th Dimensional Templar Cross constitutes one of
  The 9 CELESTIAL TRANSCIPTS from the body of inspired
  work known as **The Jain 108 Cosmometries**
  — **Rings of 24** or **Wheels** of **24**
  — The **3-6-9 Sequence**

- **PART 3:**
  — **PHI'S PALINDROMIC PRIMES (PPP)**
    7 7 5 3 1 7 1 3 5 7 7 0

PART 1

Definition of Primes and Ulam's Rose

**A prime Number can not be divided by any other number except by 1 or by itself.
They are like the atoms of creation.**

For 2,000 years, we have been told that no distinct pattern or symmetry exists within this infinite nonsense sequence!

Prof. James McCanney recently discovered waves of symmetry in the Prime Number Sequence, which means that all mathematical books need to be revised. Up till now, the military and internet encryption systems were based on the largest Prime Numbers known to us, but since this amazing discovery, NASA are trying to shut down this revelation. One of my goals is to train teachers to learn quickly how to determine the next Prime Numbers and understand the inherent symmetry and patterning in this code.

Of all numbers, the Primes are of royalty. You will see in a short while why the Queen of England wears the secret insignia for Prime Numbers which is the Prime Number Cross, what you may know of as the Maltese Cross. See Fig 1. One must wonder why this symbol has been so special over the centuries, and why it was worn on the Heart/Solar Plexus Chakra.

Currently the world daftly thinks there is no pattern in this sequence. Top professors still play around with notions like:
Why do prime numbers occur at such inconsistent intervals? Could there be one single formula that predicts all prime numbers, or something that can be said that is true for all prime numbers? Is there any kind of regularity in the appearance of primes?
Of course "They" know there is a pattern, they just don't want you to know it, that is why all the mathematics books are disharmonic or in error, and why children shut down disappointed that their insipid factory style of maths has been watered down and is no fun at all.

Fig 1
The Prime Number Cross (Maltese Cross)
worn by members of the Order Of Saint John
circa the 11th and 12th Centuries in the Holy Land.

There are 3 Phi Prime Connections or symmetries that will be
discussed in this article, I will list them here before we go into the

maths of how to derive the prime numbers from **Eratosthene's Sieve Method**.

1 – Ulam's Rose Spiral of Primes

2 – Prime Number Cross or Sequence based on the 24-ness of both Phi Codes 1 & 2. (Original to Jain)

3 – Phi' Palindromic Primes, an original discovery linking Phi to Primes! (based on Phi Code 1).

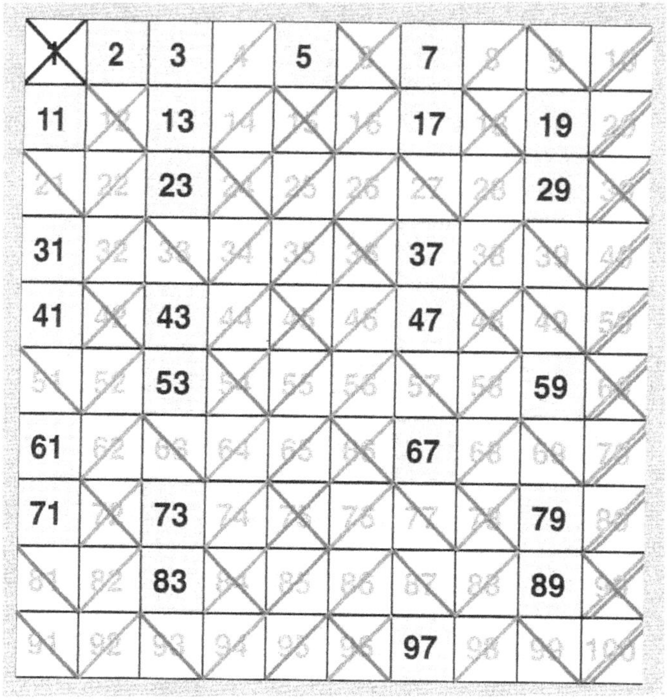

Fig 2

The Sieve of Eratosthenes created by an ancient Greek mathematician. The numbers from 1 to 100 are crossed out in certain ways until only the indivisible Primes remain.

In mathematics, the **Sieve of Eratosthenes** is a simple, ancient algorithm for finding all prime numbers up to a specified integer. Consider a contiguous list of numbers from two to some maximum number like 100.

Strike off all multiples of 2 greater than 2 from the list.

The next lowest, uncrossed off number in the list is a prime number.

Strike off all multiples of 3, then 4 then 5 etc until all the numbers remaining in the list are prime.

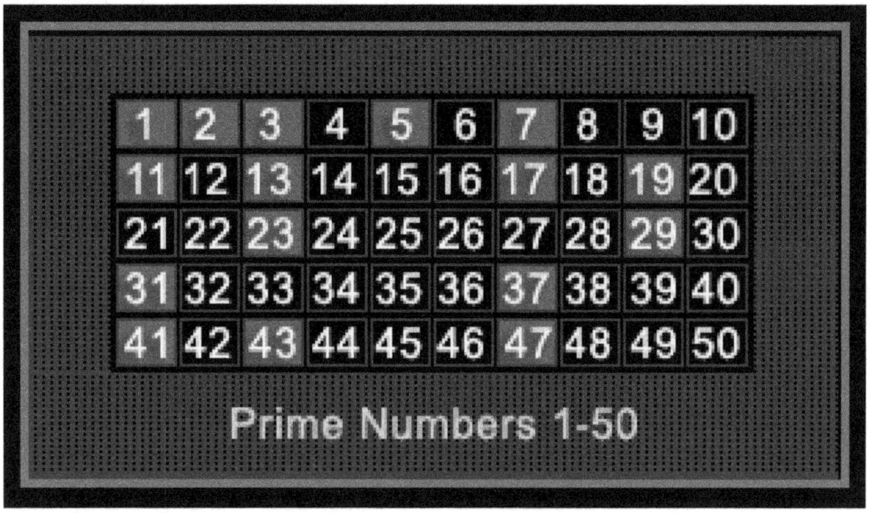

Fig 3
The list of Prime Numbers from 1 to 50

ULAM'S ROSE

Before we show how the Prime Number Cross is extruded out of the 24-ness obedient to Phi's mystery patterns, here is yet another famous pattern found in the so called non-sense symmetry of the prime number sequence. This one is dedicated to Stanislav Ulam.

Fig 4
STANISLAV ULAM who found an organic form in the infinite web of Prime Numbers, now known as Ulam's Rose

No one could supply the world with an answer and primes were believed to occur randomly. The excitement about primes flared up even more in the wake of boredom of a devoted 20th Century math-magician named Stanislav Ulam.

He put down the number 1 as the bright shining center of a universe of numbers that Big Banged outwardly in a spiral:

73	74	75	76	77	78	79	80	81
72	43	44	45	46	47	48	49	50
71	42	21	22	23	24	25	26	51
70	41	20	7	8	9	10	27	52
69	40	19	6	1	2	11	28	53
68	39	18	5	4	3	12	29	54
67	38	17	16	15	14	13	30	55
66	37	36	35	34	33	32	31	56
65	64	63	62	61	60	59	58	57

Fig 5
Ulam's Spiral of Numbers beginning from the center 1 and circling this forever clockwise.

To see this concept of spiralling numbers issuing forth form a divine centre, here is a simple spiral path that you will recognize, almost like a labyrinth path, a journey from the centre then outwards, and vice-versa.

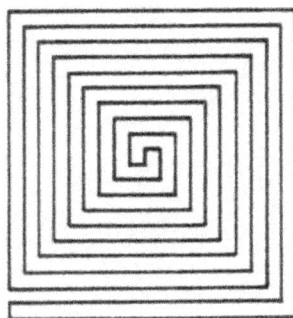

Fig 5a

Ulam's Spiral seen as a spiral pathway

This is what he thought: What would happen if we were to write the natural counting sequence of numbers, that is the consecutive numbers as in 1 - 2 - 3 - 4 - 5 - 6 - etc upon graph paper or gridded paper, in a spiral fashion, starting from the number 1 in the centre, and radiating outwards to infinity?

100	99	98	97	96	95	94	93	92	91
65	64	63	62	61	60	59	58	57	90
66	37	36	35	34	33	32	31	56	89
67	38	17	16	15	14	13	30	55	88
68	39	18	5	4	3	12	29	54	87
69	40	19	6	1	2	11	28	53	86
70	41	20	7	8	9	10	27	52	85
71	42	21	22	23	24	25	26	51	84
72	43	44	45	46	47	48	49	50	83
73	74	75	76	77	78	79	80	81	82

Fig 5b

Ulam's Spiral using numbers from 1 to 100, seen as a spiral pathway marking all the Prime Numbers with a Cross.

Then, since we are mathematical explorers here, or pattern hunters, we begin to mark in with a cross or colour in all the known prime numbers. Would we expect to see a pattern, as we approach infinity?

You can see in Fig 5b that already there is a sense of a pattern emerging, enough to want to continue with larger numbers.

Much to his amazement the prime numbers appeared to gravitate towards diagonal lines emanating from the central, see Fig 5c.

Yet there was no apparent rule that forced *all* prime numbers upon a diagonal line like that. Most of them sat on or in the vicinity of a diagonal, but some obviously didn't. Ulam ran home and expanded the spiral to cover a much larger portion of the number sequence. The strange pattern persisted. Primes had a tendency to occur in clusters and all clusters tended to make a beautiful image that could not be predicted.

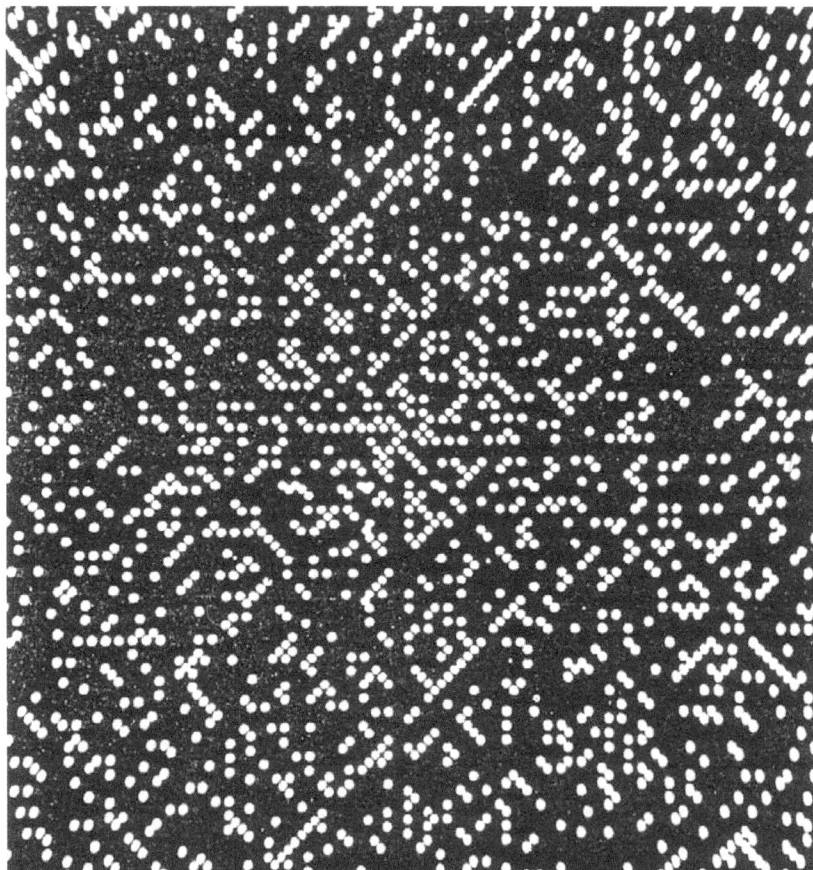

Fig 5c
Ulam's Spiral using numbers from 1 to 10,000.

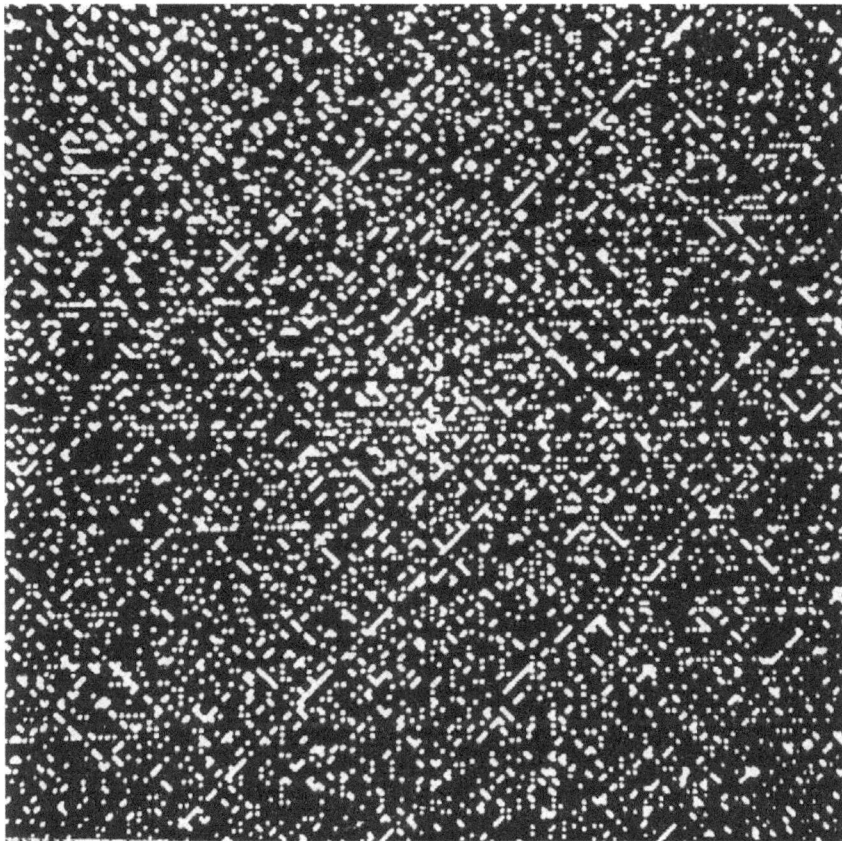

Fig 5d
Ulam's Spiral using numbers from 1 to 65,000.

It looks like something out of nature but in fact it's the prime numbers from 1 to 262,144. Like water molecules huddle together to make a snow flake according to some basic design, prime numbers huddle together to make the Ulam Rose.
The **arythmophilic** world responded in awe. There's true, unpredictable randomness in the prime number sequence! Numbers are as beautiful as nature! And up to this day every book on popular mathematics uses the word *random* in direct relation to prime number distribution.

Again, Ulam's Rose is another feather in Sri Technology's hat, for without the power of the silicon chip, we could not have had the computers to extract this amazing pattern out of the void of infinity, seen above as Fig 5d and below as Fig 5e.

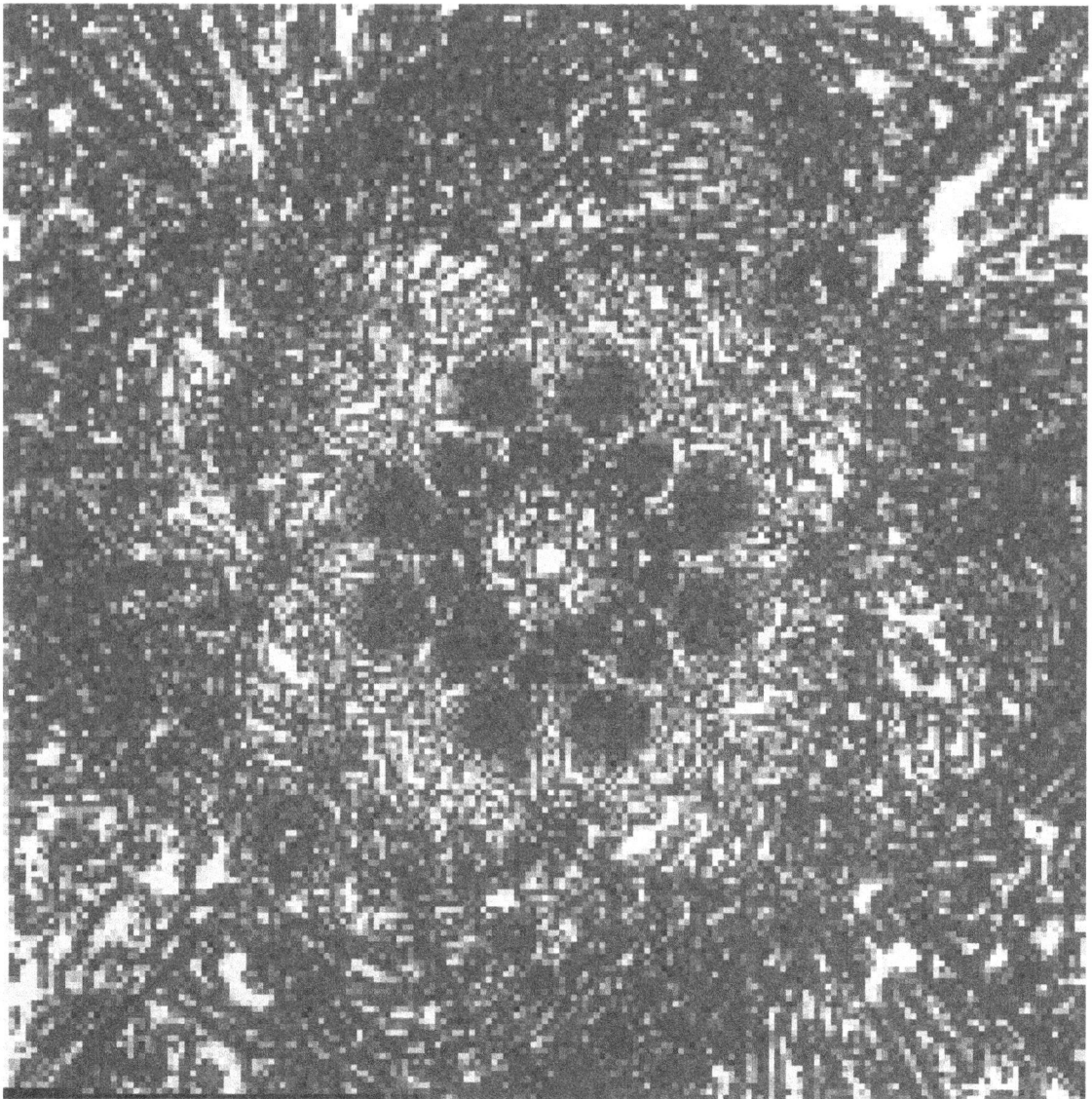

Fig 5e
Ulam's Rose of Prime Numbers:
is not a random pattern but is a flower!

[Credits: The Ulam Rose of 1 => 262,144 used here is an embellishment of an image originally created by Jean-François Colonna ©1996, CNET and the École Polytechnique, Paris France. The picture used here comes out of *Cracking the Bible Code* by Jeffrey Satinover, M.D.]

Fig 5f
The True Rose,
Ultimate symbol of Sacred Gaiaometry
that demonstrates Perfect Embedding,
Recursive Divine Proportioned Enfolding,
Self-Organized and Shareable
Non-Destructive and Forever
the art of how to Get Fractal
the art to feel the Tingle of your Fingertip
the Living Curvation
the Mathematics of Beauty
the Bliss in a Child's Eye

Here is a bit more theory on how Prime Numbers arrange themselves. Notice the 6 columns!

The Distribution of Primes in columns of 6 $(6n + 1)$ or $(6n - 1)$				
	2	3		5
7				11
13				17
19				23
				29
31				
37				41
43				47
				53
				59
61				
67				71
73				
79				83
				89
97				101
103				107

Fig 6

The Distribution of Primes in columns of 6
$(6n + 1)$ or $(6n - 1)$

This table is well known to most mathematicians. They recognize these 6 columns that harbour the prime numbers, but because they could not think circularly, as in the **Wheel of 24**, they could not grok or perceive the immense symmetry. Symmetry is the charge, is what we are seeking, and therefore teaching.

PART 2

PRIME NUMBER CROSS
GENERATED FROM CONCENTRIC RINGS OF 24!

Why have we been told for 2,000 years that there is no distinct patterning to the infinite sequence of Prime Numbers.

Yet observe what happens when you write the natural counting numbers in concentric rings of 24, suddenly there opens up a 4th Dimensional Cross, known by the ancient Egyptians and used by the Knights Templars who wore this on their breastplate or heart chakra.

Why is this number 24 important in unlocking the keys to Time Travel (24 hours to the day!) or the Physics of Time Bending.

▲ *Four salmon tails in the arms of Simo in Lapland form a cross for the see of Uppsala.*

(4 Salmon Tails, arranged into a Maltese Cross symbol, Lapland)

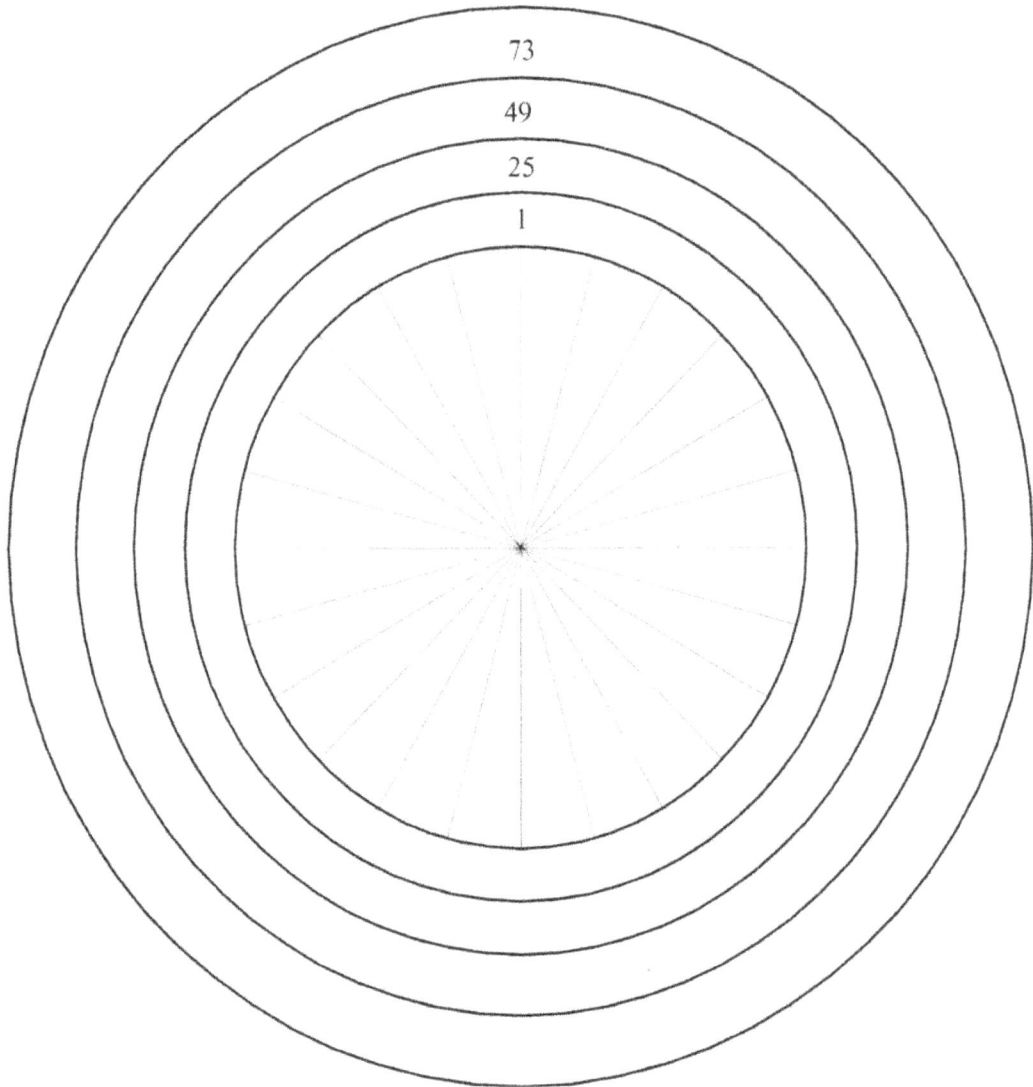

Fig 7
Worksheet with concentric rings of 24 divisions.

Upon Fig 7 you are required to photocopy first, then write in the sets of 24 numbers, in natural counting order beginning from 1-2-3-4-5-6-7 etc up to 24 in the 1st ring, then in the 2nd concentric ring the numbers from 25 to 48, and then in the 3rd ring the numbers 49 to 72 etc.

Then examine each ring, and circle all the numbers that you now know are prime to have a diagram looking something like this:

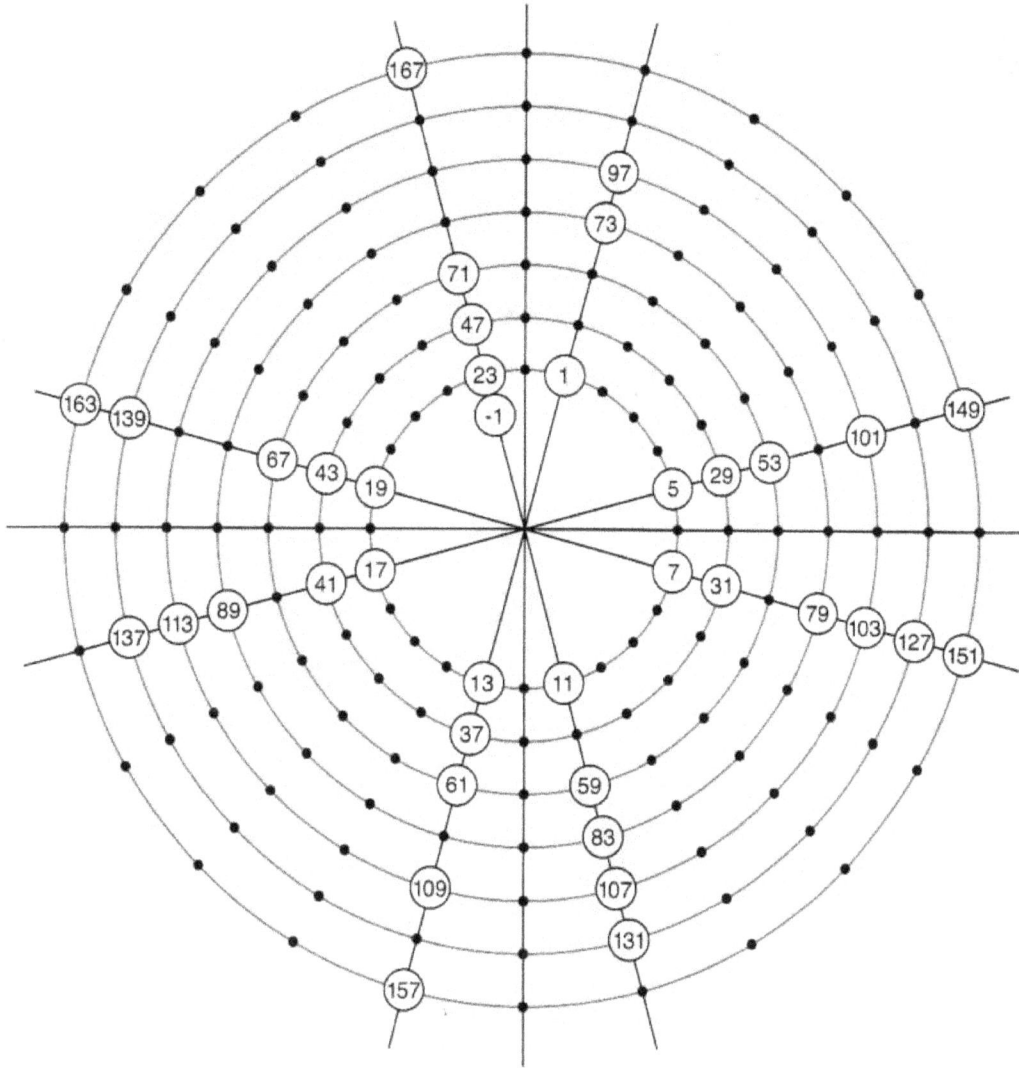

Fig 8
All the Prime Numbers have been circled
in the first 7 concentric rings of 24 divisions.

Here is another version of the same diagram, from the work of Peter Plichta a German bio-chemist who claims that this was ancient Egyptian knowledge.

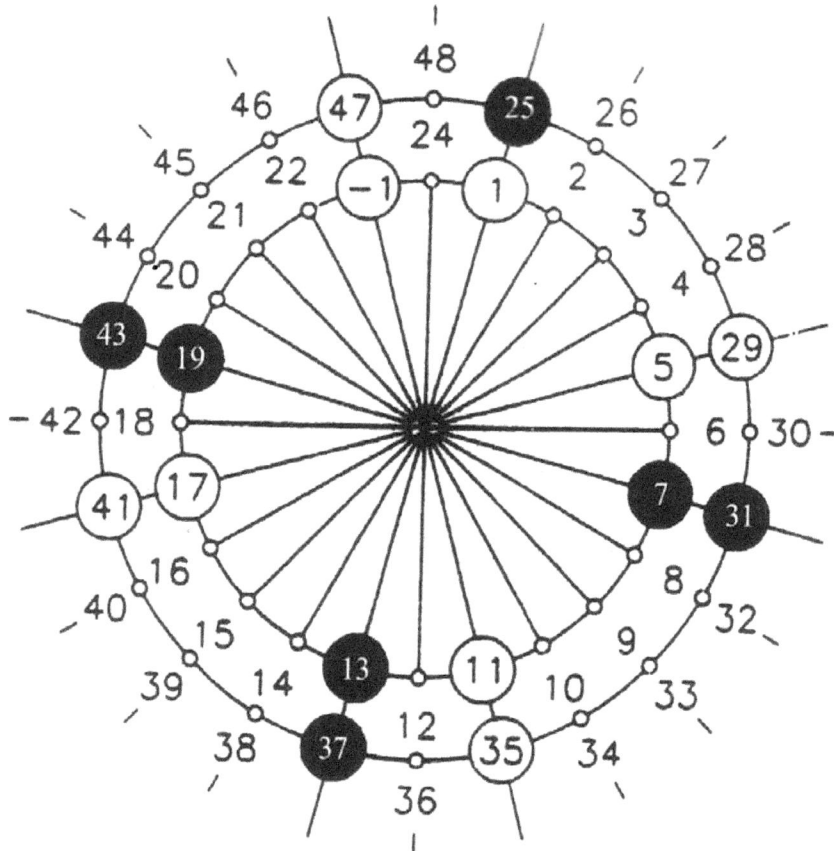

Fig 9
Peter Plichta's version of the Prime Number Cross

Do you notice that all the Prime Numbers appear to be aligning themselves along 4 distinct axes or compass points.
This could only happen by dividing the circle into 24, not any other number. The clue was given before in Fig 6 where I tabled the The Distribution of Primes in 6 columns of (6n + 1) or (6n − 1). Since 6 is a factor of 24, it raised 24-ness to being the grand secret to open up this mystery.
In the next diagram I will connect these 4 diagonals or axes where the Prime Numbers lie, and investigate the pattern that emerges:

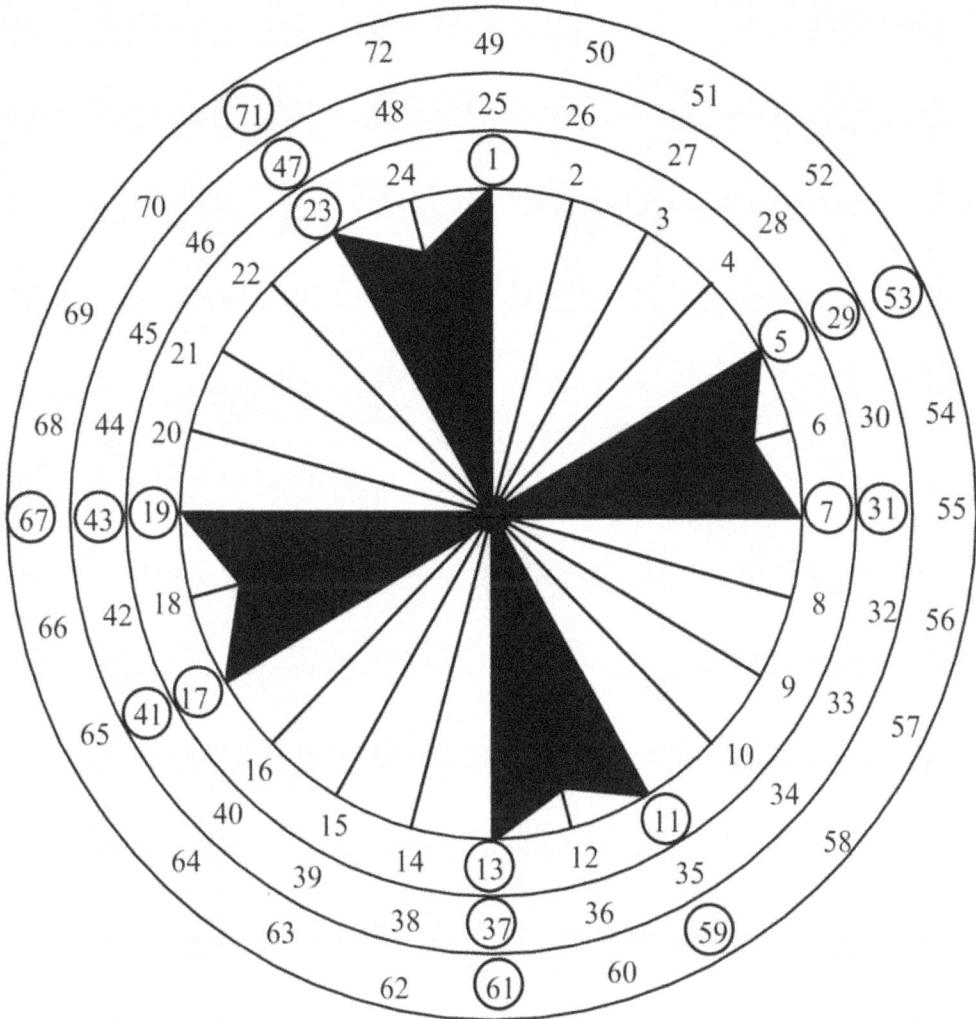

Fig 10
The Prime Number Cross
aka the 4th Dimensional Templar's Cross worn on the Heart.
Based on the division of the circle into 24
(the IcosaTetraGon = having 24 sides)

Below, the same cross, but highlighting the original colour coding in red, but shown below in black:

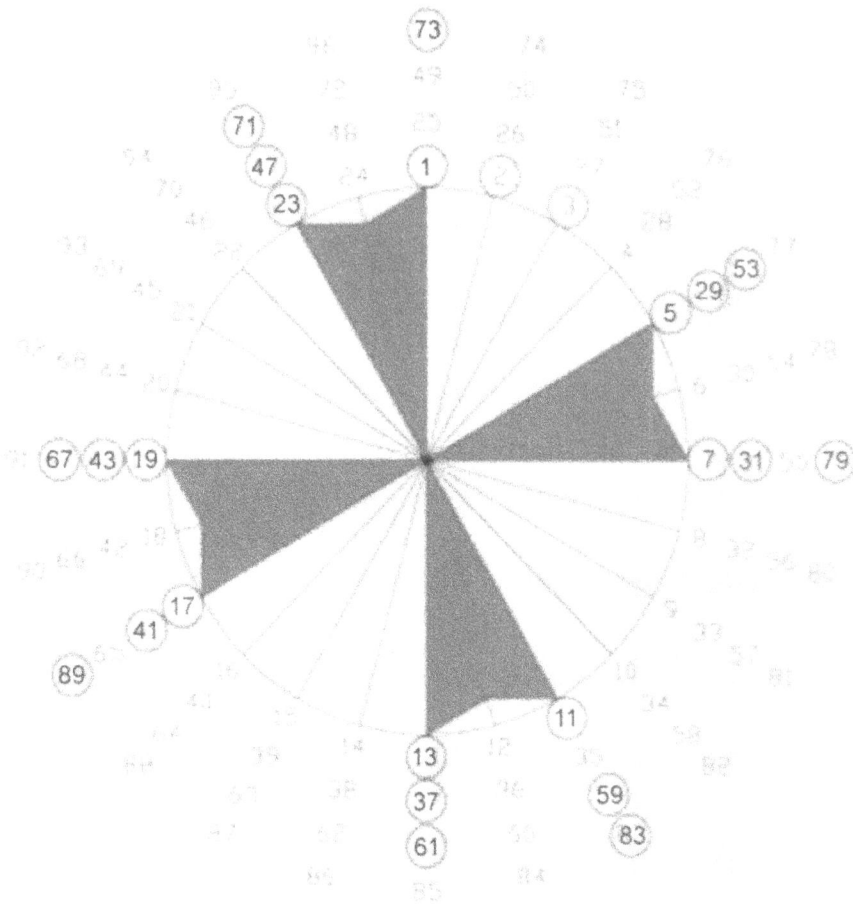

Fig 10a
Traditionally the Prime Number Cross was painted red

Historians speculate about the choice of red colouring, perhaps for the blood that was lost when the Order of St John, Crusaders and other Masonic-like groups were persecuted for this knowledge around the 11th and 12th Centuries. I personally subscribe it Kathleen McGowan's research towards the lost red symbol of Mary Magdalene's blood-line. Read in this order: "The Expected One", "The Book Of Love" and "The Poet Prince".

Worldwide, we know that the current symbol for rescue emergency healing is the Maltese Cross form also known as St John's Cross. This is shown in Fig 10b on an ambulance vehicle here in Mullumbimby, Australia.

Fig 10b
Red Cross, symbol Emergency or Rescue Healing,
shown on an ambulance. (My life was saved by them in 1984
when a thug lunged a blade through my heart chakra)!
Spending 2 days out of my body gifted me with more
insights into these da-Vinci-like Codes:
the secret is the ability to translate Number into Art.

Only Nature's intelligent choice of 24-ness generates the divine
symmetry of the Phi Code 108 Mysteries.
This symbol is found even on small coconut islands out in the South
Pacific:

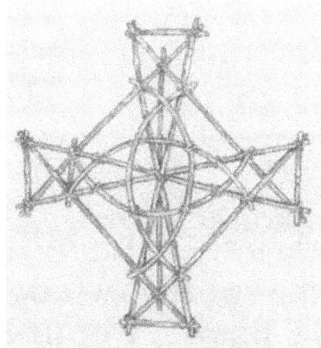

Fig 10c
Prime Number Cross Form seen in Mattang from Polynesia

Where else have you seen this ancient hidden symbol?

This portrait of Her Majesty The Queen was painted by Mr. Leonard Boden in 1968. Her Majesty is shown in the robes of the Sovereign Head of the Order of St. John and wearing the Insignia of the Order which were made for Queen Victoria. On November 28th, 1968, Her Majesty visited St John's Gate, Clerkenwell, where the portrait was on view for the first time.

Fig 11
Have a guess who wears this the 4th Dimensional Templar Cross! This ancient Knowledge of 24-ness is partly why we have 24 hours of the day etc and a Base 12 Imperial system of Measurement. Whoever has this secret, royal and confidential knowledge has the potential power to rule the world.

I found a British One Pound note and looked closely at the Queen's crown. Here a close-up of the Queen's crown, and below the actual whole note.

Fig 11a
Prime Number Cross visible
on Queen Elizabeth 11 of England's crown

Being enchanted by such an observation, I decided to pen a drawing of this. In Fib 11b below, I have typed out the hand-written writing at the base of the drawing.

"The Queen of England Adorned with the Prime Number Cross" art by JAIN 20-8-08 Tanna Island. Copied from a One Pound note that has Sir Isaac Newton on back. Nb: these 2 crosses are on her Crown (Chakra). Why? She rules the World as she understands the ancient mathematical mysteries. Nb: also there are 3 spirals like Sixes = 666 and below are 18 beads in her hair spaced every six beads = 6+6+6. She is thus Empowered.

Fig 11b

"The Queen of England adorned with the Prime Number Cross" on her crown. Art by Jain 20-8-2008, Tanna Island, Vanuatu. Copied from a One Pound note that has Sir Isaac Newton on the reverse side. Nb: these 2 crosses are on her Crown Chakra! Why? She rules this Material World as she understands the ancient mathematical mysteries. Nb: also there are 3 spirals like Sixes = 666 and below are 18 beads in her hair spaced every six beads = 6+6+6. She is thus Empowered by the Harmonics of Creation.

Just for the record, you will see this symbol everywhere. I am including a few more photos of Kaiser Wilhelm 2nd of Germany.

Fig 12
Kaiser Wilhelm 11 of Germany
wearing his royal Prime Number Cross

Fig 12a
Kaiser Wilhelm 11 of Germany
wearing his royal Prime Number Cross in full regalia.
The Maltese Cross is visible on his Heart chakra!

Fig 12b
Kaiser Wilhelm 2nd of Germany
wearing his royal Prime Number Cross in full regalia.

Germany (1867–1919).

Fig 12c
German Flag 1867 to 1919
(this image and Figs 12d,e are taken from "Flags and Heraldry")

▼ *Below German imperial flags: first row 1871-1890 emperor's standard, empress' standard, crown prince's standard; second row 1890-1918 emperor's standard, empress' standard, crown prince's standard*

Fig 12d
German Imperial Flag (left) and
flag for the Prussian Crown Prince (on right)

Germany (1937–1945).

Fig 12e
German flag with combined symbols of the Swaztika and Maltese Cross used in World War 2

H.M. THE QUEEN, ACCOMPANIED BY H.R.H. THE GRAND PRIOR, AT THE ROYAL REVIEW, HYDE PARK, 1956.

Fig 13
The Queen & the Grand Prior, 1956,
wearing proudly the symbol on his jacket.

RINGS OF 24
AND THE DERIVATION OF THE 3-6-9 SEQUENCE.

In this book (in the section on how to create the Enneagram from the Phi Code) you will have seen how the **3 – 6 – 9 Code** was generated again using our tool or weapon of Digital Compression from this Wheel of 24.

24 is the way in, the door, the StarGate!

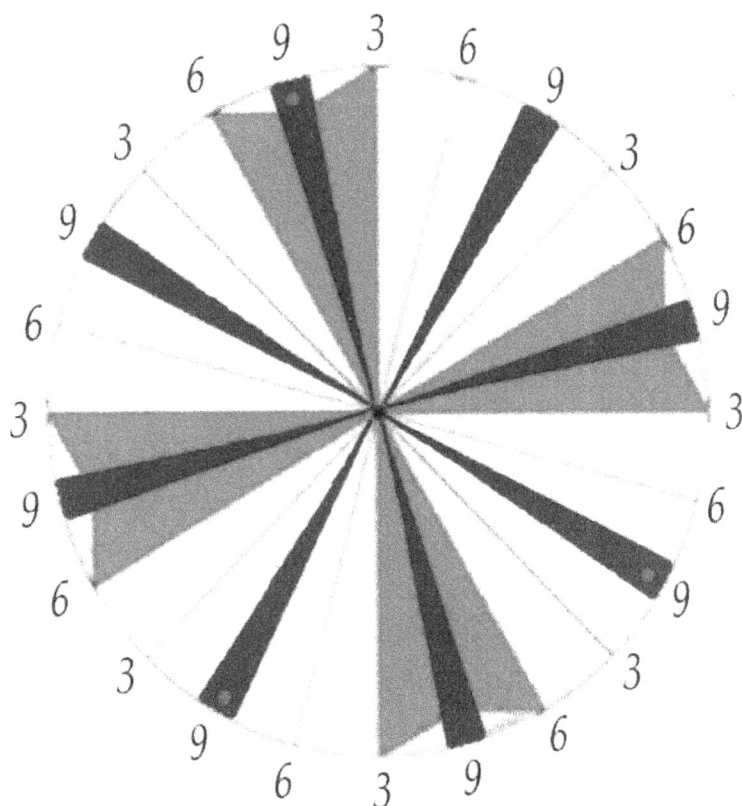

Fig 14

**The Infinitely cycling 3-6-9 Sequence
creating the Harmonics of the
4th Dimensional Templar Cross,**

(Taken from "Numbers Of Light" by Jason O'Hara, 2007)
To see how this is created,
see the article in my "The Book Of Phi" volume 5 called:
"Derivation Of The Enneagram From The Phi Code 108"
**[Nb: In my next phi book, Book 5, there will be a whole chapter on 3-6-9
as the digital compression of the Wheel of 24, and how
it sums to 144 Speed of Light Harmonic!]**

The QUEEN of ENGLAND OWNS SHRI 108!
Vs
BHARATI KRSHNA TIRTHAJI

The Queen of England possesses the 108 code!
Sorry to say this, but my mathematical research has indicated clearly to me that She and her Kingdom and Ancestry have captured, tamed and harnessed the power of the wild card known as 108. I don't mean to be disrespectful to any Vedic scholars who claim nebulous copyright on this Shri 108 Holy of Holies, the truth is that it is currently owned by the Engines of War and the Moghuls of Capitalism who surgically removed this 108 frequency from the bowels of the Phi's Powers in the Multi-Dimensions "Phi^n" that which I call the "PHI-NEST" as 108 allows non-destructive travel from the atom to the universe, the small micro picture to the large macro picture, it travels recursively, knows how to embed in Pairs fractally, is coherent and is totally self-organized. 108 is the Holy Grail, that is why this harmonic maths is not taught in schools as it deserves to be.
Yes on the superficial surface the Eastern Indians mindlessly bow unto statutes with 108 beads, shrines built with 108 bricks, chant to 108 goddesses, yet not one of them could tell you why 108 is of grand importance, and on the other side of the wide, in the hidden palatial dungeons of conspiracy, the English dinner table is headed by the family of Rothschilds, a savvy plutocracy who have mindfully, awarely, secretively, acutely and masterfully orchestrated world dominion over lesser nations, they who have won Crusade battles and terrorized temples whose invisible flag is and was 108. Yes, they who have absorbed it, owned it and militarized it.
But the world Journey is interesting now, for the Shri 108 mathematics can not be fenced or boundaried. The best that we can do to reclaim it as our own, is to teach this Jain 108 Mathemagics to students globally, not just the Indian Population, and have it out to share and reprogramme. Perhaps a few enlightened souls will take it to the zenith to equalize the power war struggles. As the next generation start using these codes that

amplify our focussed intents, the people who are fighting mindlessly might ask the question to themselves, one day, "Why am I fighting? Who am I fighting?" You see, the 108 Codes are based on Power and Harmony (The Powers of Phi, squared, cubed to the fourth power etc, and Phi is nothing more than the Mathematics of Beauty). Its that simple. Change is not gonna happen, it already is happening, even while the child is joining the dots in their studies of the Art of Number, they are actually rewiring their own optic nerves and neural pathways so that more Light quotient can come in to their Caves of Brahma, and act as coherent laser light, to penetrate the darkness of illusion.

Many parents of home-schoolers ask me if the Rapid Mental Calculation aka Vedic Mathematics is the key topic for their young teenage children to learn. It is, but it is only a tenth of the pie to eat, there are so many more other essential topics like magic squares, platonic solids, phi, to absorb.

In the bible of "Vedic Mathematics" on pages 347-348, a now famous book that has spawned a 1,000 websites, written by the spiritual head of India, Bharati Krsna Tirthaji in 1958, he actually concludes that the 15 other books on mental calculation that he wrote and that got stolen by the Vatican, were topics on astronomy, 3-dimensional solid geometry and why there can only be five regular Polyhedrons, conics, sines, calculus, the value of pi recorded sonically, and quite a long list of other important topics.

I believe that your child can literally absorb all this material, when taught correctly. I believe also, that I am the ambassador or teacher who will fulfill Bharati Krsna Tirthaji's lost curriculum, I have been writing books on most of these subjects now for 30 years. As expressed earlier in Fig 10b, the caption said: "Red Cross, symbol Emergency or Rescue Healing, shown on an ambulance. (My life was saved by them in 1984 when a thug lunged a blade through my heart chakra)! Spending 2 days "Out of my Body" gifted me with more insights into these da-Vinci-like Codes: the secret is the ability to translate Number into Art".

There are now hundreds of excellent Vedic Mathematicians, but like I said, it is only a fraction of this new bi-millennium curriculum that I have penned and ready to launch. Without boasting, I gave to the world the first complete dvd on rapid mental calculation, literally beating the Indian scholars toward this end, because I have done the research, even though I am an Arab living in Australia, teaching in the USA, my soul is from India (my Grandmother's name, from my father's side, was "Sita Hind" a very Indian title). I have fasted extensively and have been a funnel for this lost knowledge, it is not important whether or not I am an Indian scholar, means nothing, its more about the purity of intent. But the time has to be right, and the angels working with us. Even the master Bharati Krshna Tirthaji was not able to fulfil his mission in 1958, died in 1960, he has had to wait till now for individuals like Kenneth Williams (who popularized the Vedic Maths back to the Indians and wrote many books) and myself (who created the first popular article on Vedic Maths in Nexus Magazine, and created the first comprehensive dvd in the world on Vedic Maths) to pump it out; the American government suppressed all of this knowledge in the 1960's because they feared creating a race of geniuses who might ask intelligent questions.

The Author
Jagadguru Śaṅkarācārya
Śrī Bhāratī Kṛṣṇa Tīrthajī Mahārāja
(1884—1960)

Fig 15
Bharati Krshna Tirthaji (1884-1960)
The spiritual head of India
who wrote the classic
"Vedic Mathematics"

Prime Numbers: Idiot SAVANTS

"Normal people" can not give answers to the question if a big number (eg: 6,536,542,534,521) is prime without computer power in reasonable time. Autistic people are a group of human beings who are mostly unable to read and write, yet they have unlimited access to specific, accurate knowledge in the fields of mathematics, music, and other areas. This group of individuals do not acquire knowledge by learning as the average human being does. They mysteriously 'know' explicit and correct information. One may ask: "How do autistic people know certain information or possess certain skills?"

By whatever means they obtain this information, they undermine current definitions about intelligence. Does their knowledge show that a source of intelligence exist? Is it possible to tap into this source and not know of its existence?

Dustin Hoffman made autism famous in the Hollywood movie "**Rain Man**." He played the role of a mathematical genius, able to keep track of cards at casinos, yet unable to go to the bathroom alone or to make simple decisions about what clothes to wear or what food to eat. Modern science cannot explain this phenomena.

The physician, **Oliver Sacks** describes in his book:
"**The Man Who Mistook His Wife For A Hat** " a story about the twins John and Michael, who were able to define prime numbers up to 20 digits!!! very quickly. They had the biggest difficulties with the simplest additions and subtractions. Divisions and multiplications were impossible. They said,
"we can see these prime numbers!!"
This is the right brain, visual, creative, intuitive, the same realm that children sit in when they are engaged in turning numbers into art, the main principle of all that I teach: The Art of Number.

PART 3

PHI'S PALINDROMIC PRIMES

PPP

PHI CODE OF PRIMES (PALINDROMIC)

7 7 5 3 1 7 1 3 5 7 7 0

Introduction via a letter I received.

Dear Jain 108,
I am a very rich man,
aging in illness.
There is one thing I would like to know,
b4 my time is up
and I would offer you a million dollars,
what the heck!
(also being a long time Fibonatic like yourself)
for confirmation and proof
that there exists a distinct connection
between Phi (or the Fibonacci Numbers) and the Prime
Numbers.
I suspect there is.
Please contact me when you have discovered this.
I found you on your highly informative website
www.jainmathemagics.com

This was an email I received on Tanna Island from "**Greggori**" on the 8th August 2008 (08-08-08) and it made me a very rich man.

Here is the Primes in Phi, a sequence I have known and never published, and for a bit of fun, it is also a Palindrome.

I call it therefore:

PHI'S PALINDROMIC PRIMES

by JAIN 108

(nb: only tonight did I coin this title as **PPP**, which also was an island title for **P**eople's **P**rogressive **P**arty!)
written by hand 14-08-08 on Tanna Island, where I was visiting for the month and teaching students Rapid Mental Calculation, (and typed up 24-09-08 when I was back home in Oz).

How is this Palindromic (in the Greek language, literally "running backwards" or reversed) Sequence derived?
We begin with the famous ancient Pattern of Phi Code 1: the Phi Code of 24 Repeating Pattern based on the digital compression of the fibonacci series:

1 1 2 3 5 8 4 3 7 1 8 9 8 8 7 6 4 1 5 6 2 8 1 9

and rearrange it into the familiar two rows of 12 digits:

1 1 2 3 5 8 4 3 7 1 8 9
8 8 7 6 4 1 5 6 2 8 1 9

and then examine the 12 Pairs of 9.
We then investigate the differences between the two rows, whether it be we subtract the Lower Row of 12 from the Higher Row of 12 or vice versa.
Either way, we end up with a Sequence of 12 digits that reads the same as it does backwards:

(8-1)	(8-1)	(7-2)	(6-3)	(5-4)	(8-1)	(5-4)	(6-3)	(7-2)	(8-1)	(8-1)	(9-9)
7	**7**	**5**	**3**	**1**	**7**	**1**	**3**	**5**	**7**	**7**	**0**

Fig 16

The Sequence of Prime Numbers hidden in the Phi Code 108. It is now called "The Phi Prime Connection".

So here it is, a simple sequence of prime numbers based on Nature's Fibonacci Numbers, worth a million dollars.

While on the topic of the PHI CODE
there are other parts of this that we can examine, just to see if there are any other patterns.
What follows is not as important as the above, it's just for the record. You don't need to read the next 2 pages, it's another trial and error in pattern hunting.

What would happen if we multiply the 12 Pairs of Digits in the Phi Code:

```
1  1  2  3  5  8  4  3  7  1  8  9
8  8  7  6  4  1  5  6  2  8  1  9
```

such that we had this string of numbers:
```
1x8  1x8  2x7  3x6  5x4  8x1  4x5  3x6  7x2  1x8  8x1  9x9
 8    8    14   18   20    8    20   18   14    8    8    81
```

Now what is the sum of these 12 Products?
$8+ 8 +14 +18 +20 + 8 + 20 +18 + 14 +8 +8 + 81 = $ **225**
$=15 \times 15 = $ **15²**
That's interesting, the sum of this multiplied and added sequence is the shape of a squared number! And 15 is indeed an Anointed Number, being the magic sum of the first and most simple magic square of 3x3.

Similarly, with the intent to discover, is there an interesting proportion between the sum of the top row of 12 digits of the Phi Code, compared to the bottom row of 12 digits of the Phi Code:

```
1  1  2  3  5  8  4  3  7  1  8  9
8  8  7  6  4  1  5  6  2  8  1  9
```

Adding the sum of the top row gives:
$1 + 1 + 2 + 3 + 5 + 8 + 4 + 3 + 7 + 1 + 8 + 9 =$ **48**

Adding the sum of the bottom row gives:
$8 + 8 + 7 + 6 + 4 + 1 + 5 + 6 + 2 + 8 + 1 + 9 =$ **69**

What is the ratio of the Top row to the Bottom Row?
$= 48 / 69 =$ **.695652174**
or
$= 69 / 48 =$ **1.4375**

(So we reach perhaps a dead end, but still these numbers highlighted can be indexed in the Harmonic Stairway (aka Jain's Dictionary of Anointed Numbers) in the future event that someone will find other correlations and meanings associated to these frequencies. Its not important now, its all about archiving and indexing these harmonics which is a valuable and useful exercise).

We can look also at the sum of the 12 Prime Digits Sequence:

$$7 + 7 + 5 + 3 + 1 + 7 + 1 + 3 + 5 + 7 + 7 + 0 = \textbf{53}$$

From these above investigations we can deduce that there are no more bells ringing or obvious correlations,
we can stop here and gloat at the fact that there is indeed a prime relationship hidden in the Phi Code of 24.

[**As a side note**, it was interesting on the next day, 15-08-08, I was walking early morning to the local organic market @ Tanna village, and saw a big sign "PPP" where loud music was coming from, promoting a political party "**P**eople's **P**rogressive **P**arty" written within a circle.
I thought that was very synchronistic as last night I was writing up the previous page of info and decided to coin the title "**PPP**" to mean "**Phi's Palindromic Primes**" which is circular and therefore an infinitely running or repeating sequence.
It's a special feeling knowing that I am tapped into the Collective Consciousness. The week before, I knew there were political parties canvassing their causes for the next upcoming Vanuatu elections, but I did not consciously know of any party referred to as PPP.
Jain 108   15-08-08   Tanna Island, Vanuatu].

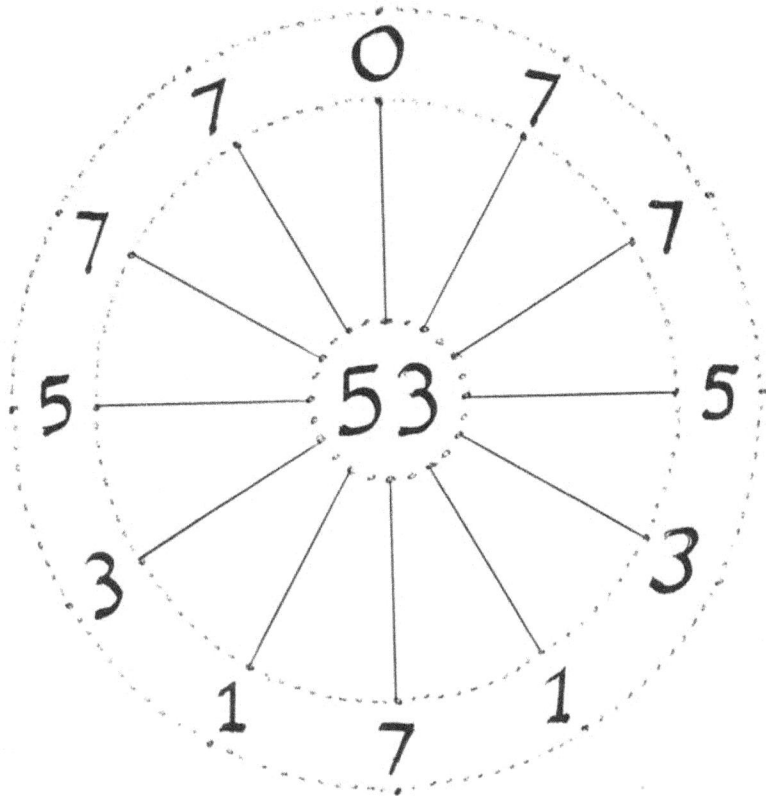

Fig 17
Diagram showing the 12 (or really 11 digits)
of Jain's Phi's Palindromic Primes

At this point we need to ask a few important questions:
Is this a necklace or an infinite wheel of 11 or 12 prime digits?
Lets examine the Phi Code 1 again and the embedded or beautifully hidden Prime Symmetry.

```
1  1  2  3  5  8  4  3  7  1  8  9
8  8  7  6  4  1  5  6  2  8  1  9
```

7 7 5 3 1 7 1 3 5 7 7 0

Notice that the above 12 digit sequence of Primes,
is really an 11 digit sequence, when not counting the Zero,
therefore we can carve this into stone:

7 7 5 3 1 7 1 3 5 7 7

Notice also that the number "11" is a Prime Number!
Notice also that the sum of these 11 digits, which is "53", is also a
Prime Number!

"God is a Sphere whose Centre is Everywhere
and Periphery nowhere".

- **Pascal**, quoting Hermes Trimegistus.

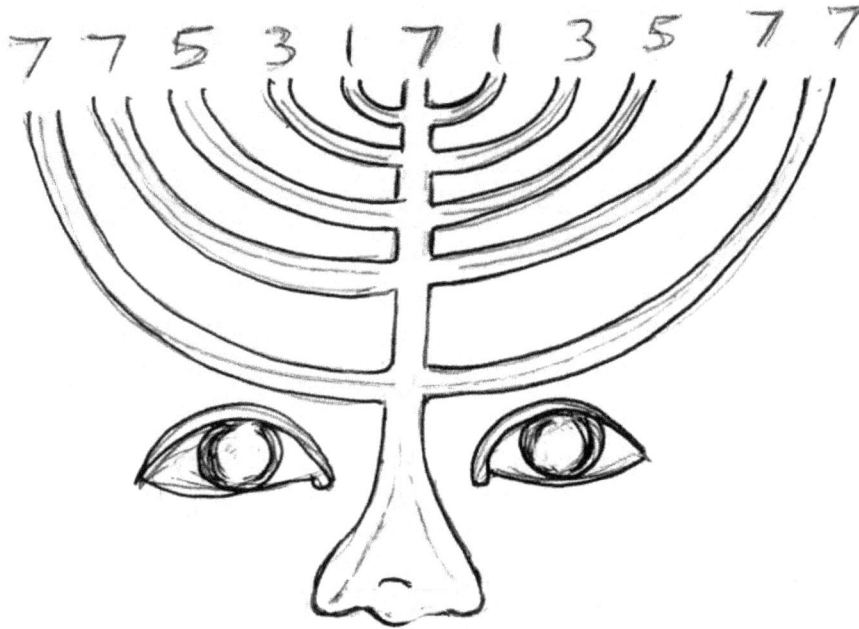

Fig 18
"The 11-Pronged Candlestick"
"The 11-Branched Tree"
that askes the embarrassing Question from Antiquity:
"IS 2 PRIME"

"IS THE NUMBER TWO PRIME?"

The Infinite Sequence of Primes:
7 7 5 3 1 7 1 3 5 7 7 0
cleverly uses all the Prime Numbers under 10.
The Number 2 is missing! Does this suggest that 2 can not be classified as Prime.
Most books teach that the Primes under 10 are 1, **2**, 3, 5, 7.
I suggest that it is not, since 2-ness is the emblem of all doubling and factoring and evenness. There are no Prime Numbers that are even numbers, and 2 is the first even number, so how can it be Prime. I put forth, from this PPP Sequence that the first Primes under 10 (written as "<10") are 1, 3, 5, 7.
This may or may not be fact, but we are here to question everything, just because it is written in books does not mean it is true. That which is demonstrated here, this new notion, is the

highest level of reasoning or intelligence based on "**INTUITIVE MATHEMATICS**". It has been an ancient problem question: "Is 2 Prime?"

Thus I put forth that the definition of Prime Numbers is more than "a number that can be divided by 1 or itself" but also that it can not be an "Even Number" like 2, 4, 6, 8 etc, or any even number that is formed by self-addition, like 4+4=8, or 3+3=6 or 2+2=4 therefore it is logical and intuitive that 1+1=2 can not be a Prime Number. So simple. No argument. In this light, "2" is the symbol of everything that is "Even" or "Doubled" and is thus the opposite of Prime.

When I say "Intuitive Mathematics" I mean that you may not immediately grok or understand what I just defined as "2" being the emblem of Evenness and is the polar opposite of Primeness. It sounds mysterious and even contradictory to what you may have been taught, it is spectral, unfathomable, but over time, chewing over such important questions, it appears to make sense. To me, it is clearly simple to understand, that 2 is not a prime number, but for you, it may take several weeks or many moons to understand this nebulous concept of Intuitive Mathematics. Or even years before the "pennies drop". One day, it will just dawn on you that 2 is not and can not be a Prime Candidate. It then becomes obvious, self-evident.

THE END

To **summarize**, how did we deduce this Phi-Prime Connection?
Via Pattern Recognition!
By engaging in the visual right brain. (Numbers or maths is left brain logical, rational, analytical; and Patterns and Pictures are right brain, visual, creative, intuitive, thus seeing numbers as art joins both hemispheres and creates whole brain holographic learning). We looked at the living mathematics of nature, compressed the infinite fibonacci series to a distinct cycle of 24 repeating digits (whose sum is 108), we restructured the one long line of 24 to sit as 2 rows of 12, or 12 Pairs that sum to 9, then examined the voltage differences between these successive pairs and arrived at an 11-branched tree whose fruits were Prime Numbers:

7 7 5 3 1 7 1 3 5 7 7

an Eternal Cycle or Rhythm circulating the first and only Prime Numbers under 10 (since we live in a world based on Base 10, humans or anthropods having ten fingers and ten toes).

"Heavenly Rose" by Lily Moses

"I am not a teacher,
But an awakener".
- Robert Frost

Philosophy is written in this grand book-
I mean the universe- which stands continually open to our gaze, but it cannot be understood unless one first learns to comprehend the language and interpret the characters in which it is written.
It is written in the language of mathematics, and its characters are triangles, circles, and other geometrical figures, without which it is humanly impossible to understand a single word of it; without these, one is wandering about in a dark labyrinth.
-Galileo Galilei
Il Saggiatore [1623]

"Guardian Of The Gateway" by **Lily Moses**, 2010

CHAPTER 4
108 PHI CODE 1
as MYSTIC COGGED WHEELS
and other
CIRCULAR INTERPRETRATIONS

Includes:

- **More Translation of Number into Art**
- **PART 1: Plotting PC1 (Phi Code 1) into 24 Divisions and 9 Concentric Rings for Volumetric View**
- **PART 2: Plotting PC1 from Centre and Outwards**
- **PART 3: Plotting PC1 from Outwards towards Centre**
- **PART 4: Plotting PC1 Pairs into the Circle of 108 Segments**
- **PART 5: Plotting PC1 onto the 9 Point Circle**
- **A New Model: How to Uni-Phi**
- **Early Phi Art by Jain from the 1980s**

24-11-2007 and 13-10-2008 Mullumbimby Creek, Far North NSW, Australia

> *"You never change things by fighting the existing reality.*
> *To change something,*
> *build a new model that makes the existing model obsolete."*
>
> **- Buckminster Fuller**

Plotting PC1 into 24 Divisions of the Circle and 9 Concentric Rings for Volumetric View

We have previously plugged the 24 Repeating Pattern of the Phi Code into a 4x6 Frame and extruded harmonics of 666. We also plugged the data into a 3x8 Frame and found more hidden patterning.

Let us now explore what these same infinitely repeating 24 numbers of the Phi Code appear like when plugged into a wheel with 9 concentric rings to define in detail the teeth or the cogs of a wheel as the pattern unfolds.

Before starting this chapter, I am creating a new jargon here, coining the expressions, "PHI CODE 1" and "PHI CODE 2" to now represent the two long-winded titles:

"PHI CODE 1" = The Phi Code 108 based on the Compression of the Fibonacci Numbers, the Linear Sequence of 24 Repeating Numbers and whose Pair (9,9) is at the end of the 12 Paired Sequence. Abbreviated to "PC1".

"PHI CODE 2" = The Phi Code 108 based on the Compression of the Powers Of Phi Numbers, the Multi-Dimensional Sequence of 24 Repeating Numbers and whose Pair (9,9) is in the approximate centre of the 12 Paired Sequence. Abbreviated to "PC2".

That is why the title Phi Code 108 is now pluralized to PHI CODE**S** 108.

In this 4th book on Phi, I am not releasing the actual Phi Code 2, as I have chosen to deal only with Phi Code 1 and will reveal the amazing Phi Code in the 5th book on Phi.

There may be other undiscovered Phi Codes.

Before we start plotting the Phi Code 1 into several circular experimental versions, I believe it is important that you first learn the Phi Code 1 by heart, to your memory banks, so here, on page 127 is a worksheet that allows you to get the feel of how the numbers run:

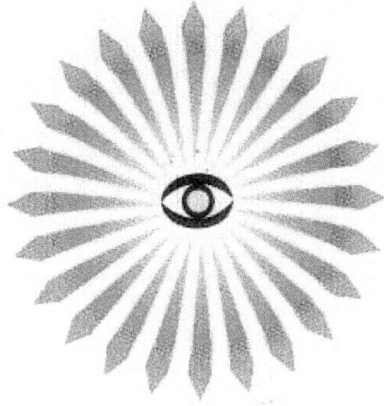

"**24 Rays**" symbolic of the 24 Repeating Pattern
in the Fibonnaci Sequence,
is really a **StarGate** or **Time Code of 24 hours** in the day.
(image taken from logo for "Inner Eye Publishers").

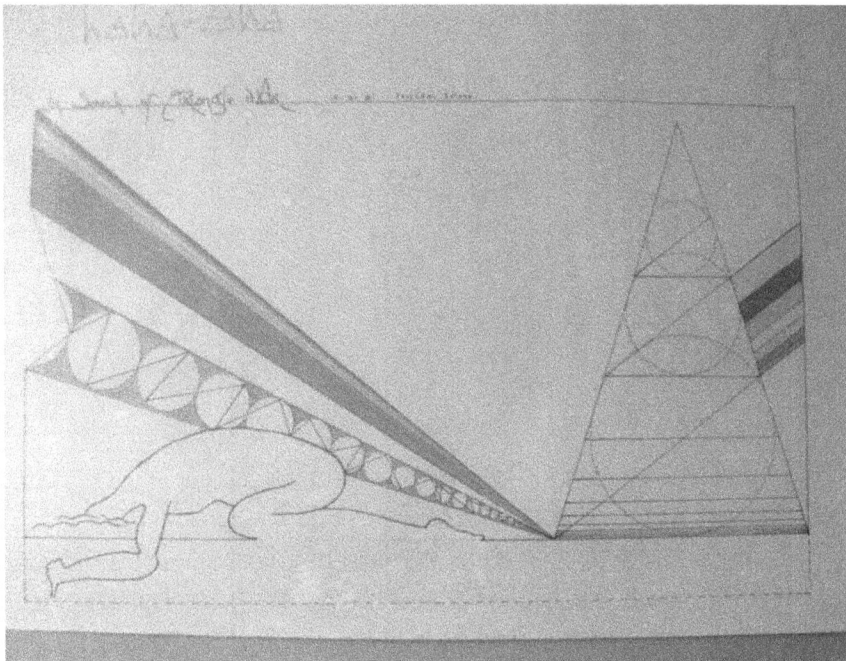

Art Of Jain 1981, entitled "IN SEARCH OF TRIANGLE D'OR".

Shown is the Golden Triangle on the right, and on the left a figure,
me, is seen burying his head in the sand. Yet light and rainbows
abound. The overall image fits perfectly into the Golden Rectangle.

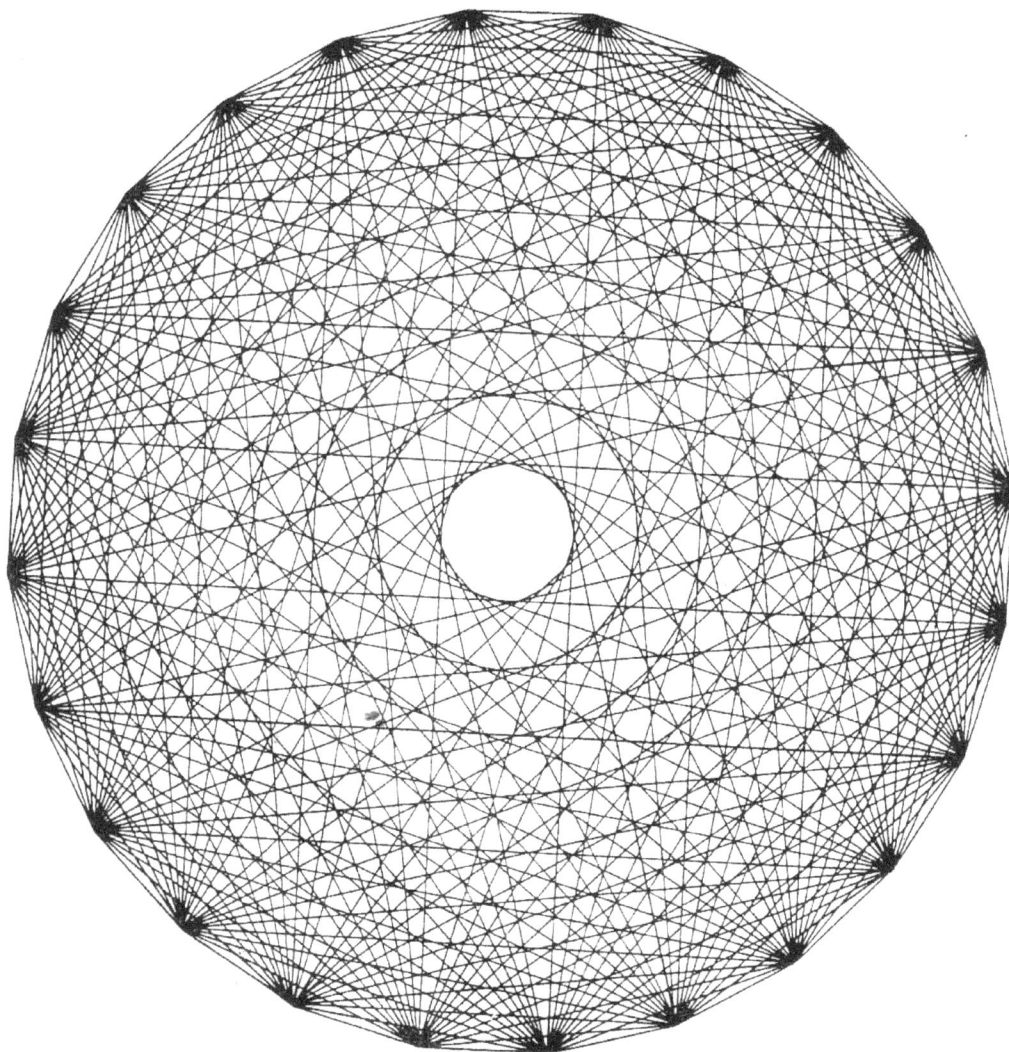

Fig 1

Worksheet: MEMORIZING PHI CODE 1

To get a feel what it is like to simply write out these 24 repeating numbers of the Phi Code 1, "Off By Heart" write these numbers from the top petal, going clockwise, and do it in a manner where you have memorized these numbers. eg: by memorizing the first set of 12 numbers of Phi Code 1, you already know that the second set of 12 numbers are the complementary pairs of 9. I find it easier to remember the first set of 12 numbers in triplets, like this:

1 1 2 – 3 5 8 – 4 3 7 – 1 8 9

(Image is a **Mystic Rose of 24 Points**: as all points are inter-connected).

To be able to understand Fig 2a, and for you to actually draw it for yourself, you are required to visualize the original wheel as a pure circle, as our starting point, with 9 concentric rings around it, and 24 radials, which will capture all the 24 numbers to be drawn as bar segments as in a bar graph or as pure arcs then shaded in as a circle segment of 1/24th. Each blackened-in arc length depends on the value given from 1 to 9, or can be seen as 24 teeth of a cog of a wheel.

Or, you can visualize the original wheel as a pure 24-sided polygon (a male shape, as in sacred geometry, any straight line is considered male, and any curved lines are considered female). Technically, the name of this shape is called a **"Icosatetragon"** (Icosa-tetra-gon, from the Greek words: "Icosa" means "20", "tetra" means "4" and "gon" means "side". This would form a shape that overall appears like a spider's web. Some books spell it as "Icosi-Tetragon".

To define the Original wheel, upon which you will draw in the cogs or wave-like undulations, whether as curved female arcs or straight male lines, I have decided, in the following worksheet, to draw in the 24 straight lines to form the Icosatetragon. As you translate the following 24 Repeating Pattern of the 108 Phi Code into this wheel, it is your choice whether you use arcs or straight lines to complete the pattern as shown in Fig 2c.

You may want to photocopy Fig 2a several times and work upon these copies, leaving Fig 2a untouched as your master copy.

To start plugging these 24 numbers into the wheel, you need to mark the top most or northern point of the wheel, in Fig 2a, call it 1; then going clockwise, the next number is also 1, then 2 then 3 then 5 then 8 etc... Its like a clock of 24 divisions. For every phi code number from 1 to 24, it will be sketched as a length, eg: the number 3 would mean you are required to colour in or sketch 3 of the 9 rings. That is how the 24 teeth of the wheel are formed. It is like a horizontally flat bar graph that is joined end to end to make a circle.

Notice, in the worksheet in Fig 2a, that the radius of the original wheel is 45mm, and surrounding it are 9 x 5mm concentric rings, meaning that the distance of the cogs is also 45mm, the same as the radius. Thus the proportion of Inner Wheel Radius to Cog Radius is "1:1".

This is a creative exercise, one based on mathematical and geometrical discovery. As a future project, you can change Fig 2a and redraw it in different ways.

eg: you can choose any proportion for the wheel-cog ratio. For example, if we call the radius of the Inner Wheel as 1 unit, we can make the distance of the cog's radius (ie: the 9 concentric rings shown as dots) to be more than one, in the ideal phi proportion of "1 : 1.618". Or we could choose the radius of the Inner Wheel as 1 unit, so that the radius of the cog is The Square Root of Phi or 1.272 which is what it takes to literally "**Square The Circle**", by being in this approximately **1:1.272** ratio or **8:9** in the Egyptians mathematical lore. We would need to do this as accurately as possible on a computer using the "Illustrator" program.

Ultimately, whatever ratio we use, the purpose of the exercise is to merely "**Translate Number Into Art**" and find practical applications of this data for our future technologies to remain in resonance with Nature's choice of numbers or natural frequencies.

[**I can see a children's playgroud**
where these 24 numbers are like a Stonehenge arrangement
where the 24 numbers are 24 volumes standing vertically,
and joined side by side circularly
so a child can climb over the various sized towers.
This must be based on the fact
that each of the 24 rectangular prisms
have a standing base of 1x1 unit
and the height will therefore depend
on the actual Phi Code Number from 1 to 9.
The Children are playing with Phi
imbued by its pleasant architectural forms
rising and falling, peaks and troughs,
yet there is a hidden & secret balance].

Jain Mathemagics Worksheet 2007
"Phi Code as a
Mystic Cogged Wheel".

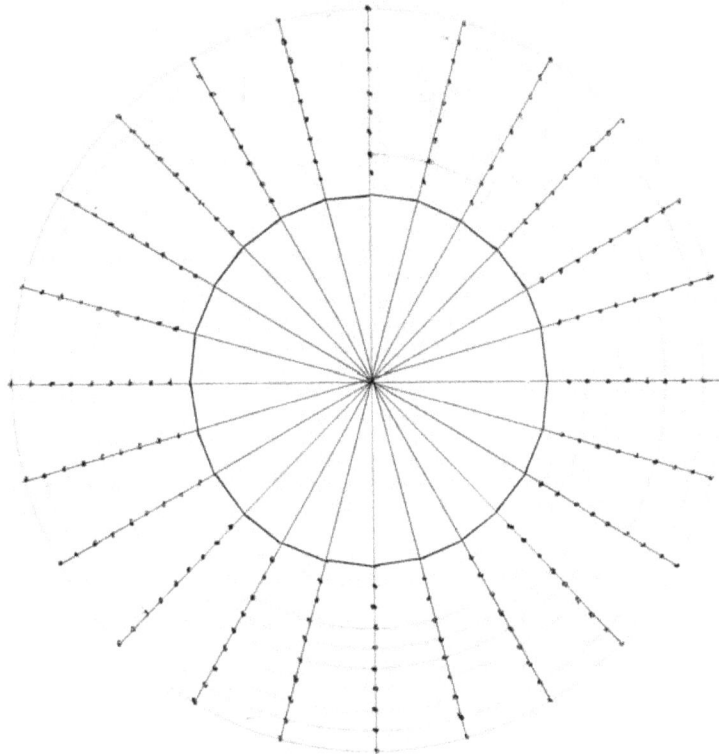

the infinitely repeating 24 Pattern. Jain 108											
1	1	2	3	5	8	4	3	7	1	8	9
8	8	7	6	4	1	5	6	2	8	1	9

Fig 2a
Worksheet for Plotting the 24 Repeating Phi Code 1
Surface Area onto a wheel or 24-sided IcosaTetraGon,
showing the Circular Cogged or Teethed Pattern.

(Arabic Tiling)

Getting the Student to draw their own Grid:

For the teacher in the classroom, who would like the student to learn how to create their own original worksheet, see Fig 2b, by dividing the circle into 24 sides, here is another worksheet that allows the student to draw in their own 9 concentric circles, more as an exercise to feel how to use a compass. Notice on the horizontal line or diameter, there are 9 dots indicating the 9 circles to be drawn. Plus this worksheet allows the student to write in the necessary information at the bottom of the page, eg: to draw Fig 2b, the 24 numbers need to be written into the boxes, as well as to be written circularly around the largest circle.

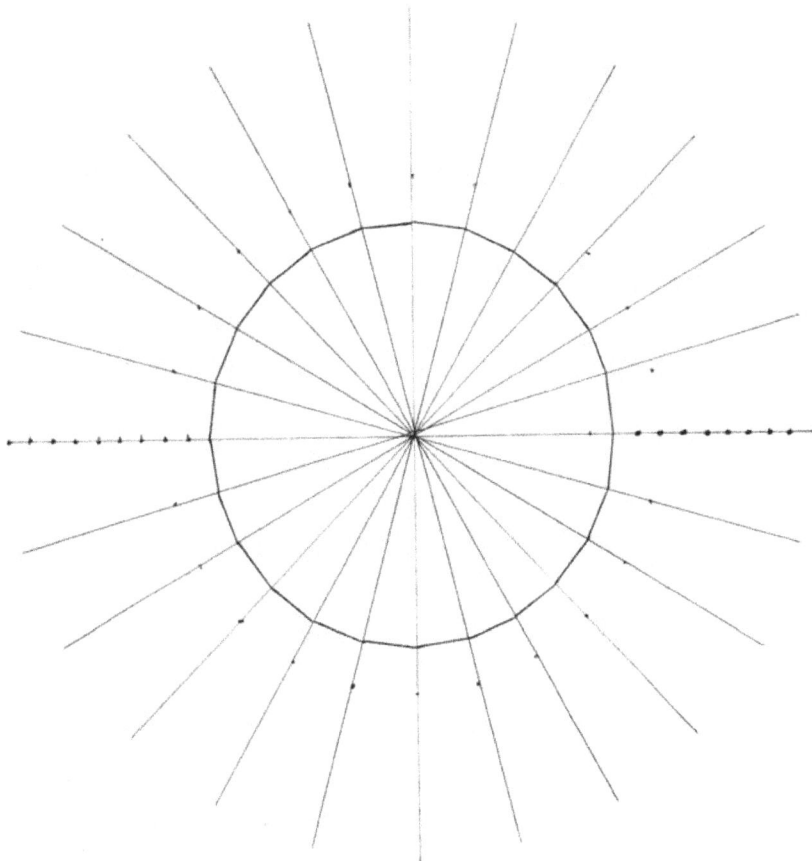

Jain Mathemagics Worksheet 2007
"Phi Code as a Mystic Cogged Wheel".

the infinitely repeating 24 Pattern. Jain 108

Fig 2b

Worksheet for Plotting the 24 Repeating Phi Code 1

onto a wheel or 24-sided IcosaTetraGon to show the Circular Cogged or Teethed Pattern.

The student is to draw their own 9 Concentric Circles.

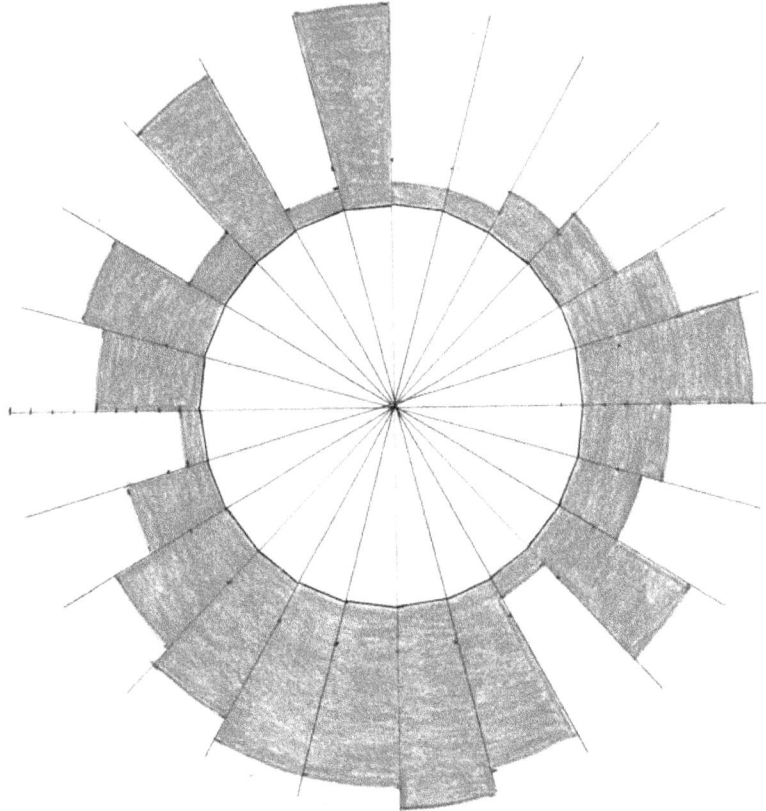

the infinitely repeating 24 Pattern.											Jain 108
1	1	2	3	5	8	4	3	7	1	8	9
8	8	7	6	4	1	5	6	2	8	1	9

Fig 2c

The Mystic Cogged Wheel, Based on Jain's Infinitely Repeating 24 Pattern of the Phi Code 1

[1-1-2-3-5-8-4-3-7-1-8-9-8-8-7-6-4-1-5-6-2-8-1-9]

Jain 108's Phi Code: an Infinitely Repeating 24 Pattern

Based on the Digital Compression of the Fibonacci Numbers

Into Single Digits. This one is termed "Linear"

aka PHI CODE 1 (in contrast to "Multi-Dimensional" aka PHI CODE 2 as will be explained in volume 5).

Fig 3a

**Phi Code 1 shown as a necklace of 24 repeating digits.
It is surrounded by 3 E-A-R-T-H s that rereads as H-E-A-R-T.
This is the origin and substance of Jain's Logo: EarthHeart.**

Ist Set of 12 Numbers	1	1	2	3	5	8	4	3	7	1	8	9
2nd Set of 12 Numbers	8	8	7	6	4	1	5	6	2	8	1	9

Fig 3b

Phi Code 1 cut in half.
The Double 9 Pair is at the end of the sequence.

The Phi Code of Complementary Pairs summing to 9 is the essence of **Base 12 Galactic Mathematics**.

Both 9 and 12 are Galactic Bases, compared to Base 10, our decimal system which allows us to perform amazing Rapid Mental Calculation (Vedic Mathematics) and is called Earth Mathematics.

Base 10 is our human correlation, since we have 10 fingers and 10 toes, it is the everyday mathematics. But when we need to build pyramids and stonehenges we work in 12x12 or 144 square inches to the square foot, since **144** is the Harmonic of Light. (Study the works of **Bruce Cathie**, the ex-New Zealand pilot, to know why 144 is equated to Light).

(A Wrought Iron sculpture displaying a wheel of 12 Rays within another larger wheel of 12 Rays, as a central motif. A very suitable Yantra or Power Diagram capable of capturing the 24 Repeating Code of Numbers).

PART 2

PLOTING PHI CODE 1 PAIRS FROM THE CENTRE OUTWARDS

JAIN · MATHEMAGICS WORKSHEET.
GRID / TEMPLATE for PLOTTING the
12 PAIRS of 9ness. (Using 10° radials)

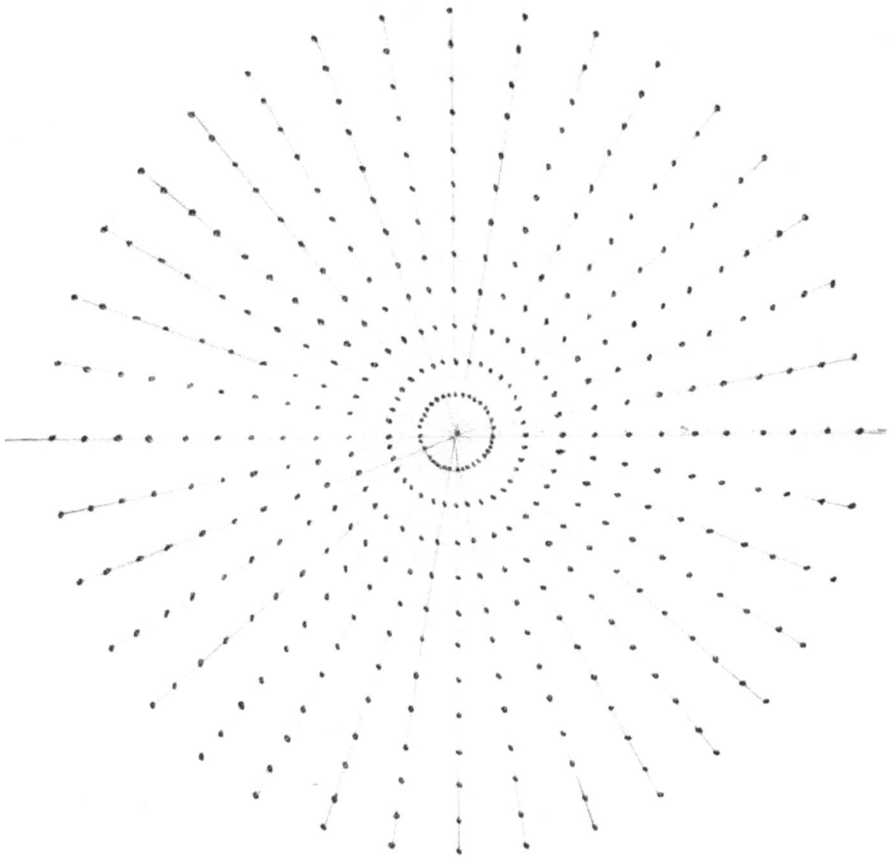

Fig 3c

Grid Template for Plotting the 12 Pairs of 9-ness, for the Phi Code 1. Use of 36 Radials. The Order of Plotting the Pairs can be geared from the Centre and Outwards or in reverse order: from the Outer Rim and working towards the Centre.

In Fig 3c above, I have given you just the original grid work of dots, achieved by dividing the circle into 10 degree radials or increments. That means dividing the circle into **36 divisions**. You can use 18 divisions of 20 degree, but to get the precise geometry that you will appreciate later when you actually draw the Pairs for yourself, you will realize that it is better to divide again and work with the 10 degree spokes.

In Fig 3d you will need to write down, in the space provided at the bottom of the worksheet, the 12 Pairs of 9 for Phi Code 1, actually you are able to remember this Pattern of numbers and not need to copy them from a written table. I do advise that the serious student commit this Phi Code 1 to memory, embed it in your Heart.

Visualize that the circle is really two joined hemispheres or semi-circles. The top or northern semi-circle of Fig 3d will carry the top half of the 12 Pairs of 9, that is, starting firstly from the centre and going outwards, you will be required to plot the numbers:

1, 1, 2, 3, 5, 8, 4, 3, 7, 1, 8, 9 .

Conversely, you will be required to plot:

8, 8, 7, 6, 4, 1, 5, 6, 2, 8, 1, 9

into the bottom half or lower or southern semi-circle.

All plotting of numbers or pairs is achieved by shading in the segments that are defined. Notice also in the next diagram of Fig 3d, I have achieved half of the project at hand, by merely outlining the 12 Pairs of 9 based on the linear Fibonacci Sequence that is boxed in at the bottom of the worksheet. Fig 3d is the same as Fig 3e. I showed Fig 3d as the halfway point of blackening in the data, in the event that this mere outline may reveal more meaning or at least show how much work is required to achieve the completed Fig 3e.

Notice also that there are 2 semi-circles of numbers ranging from zero to 9 on the outermost circle. Its like a compass, the "0" is the top of the diagram, and there are two number "1"s on the left and right of this zero northern or zenith point. Thus if you wanted to shade in a segment that was defined by "1" in the Phi Code, you can easily see the boundaries of this segment. These boundaries are created from the 10° radials, not the 20 degree radials. That is why we divided the circle into 36 divisions, not just the logical 18. Notice also that the last pair of the Phi Code, the double 9 Pair, is two semi-circles meeting which therefore creates a full circle. As a variation of this diagram, I would like to see the two hemispheres

slightly apart, by a fraction of a millimeter, to suggest that these Pairs are magnetically drawn to each other, like **Twin Flames** seeking one another's essence and divine form.
When fully sketched in, Fig 3d will look like this **StarGate**: Fig 3e

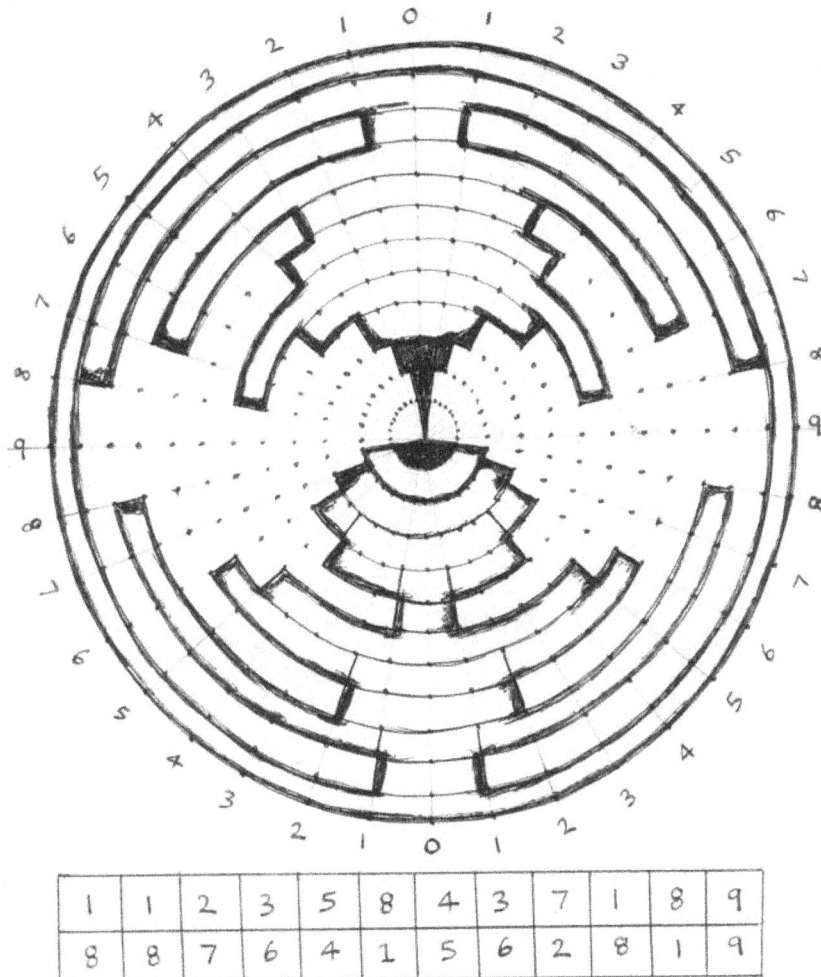

JAIN · MATHEMAGICS "WORKSHEET.
GRID / TEMPLATE for PLOTTING the
12 PAIRS of 9ness. (Using 10° radials)

1	1	2	3	5	8	4	3	7	1	8	9
8	8	7	6	4	1	5	6	2	8	1	9

Fig 3d
Showing the half-way point of construction for Phi Code 1, outlining the required segments of the 12 Pairs of 9 before shading them in completely. This gives an outline of what work is required to achieve the next diagram.

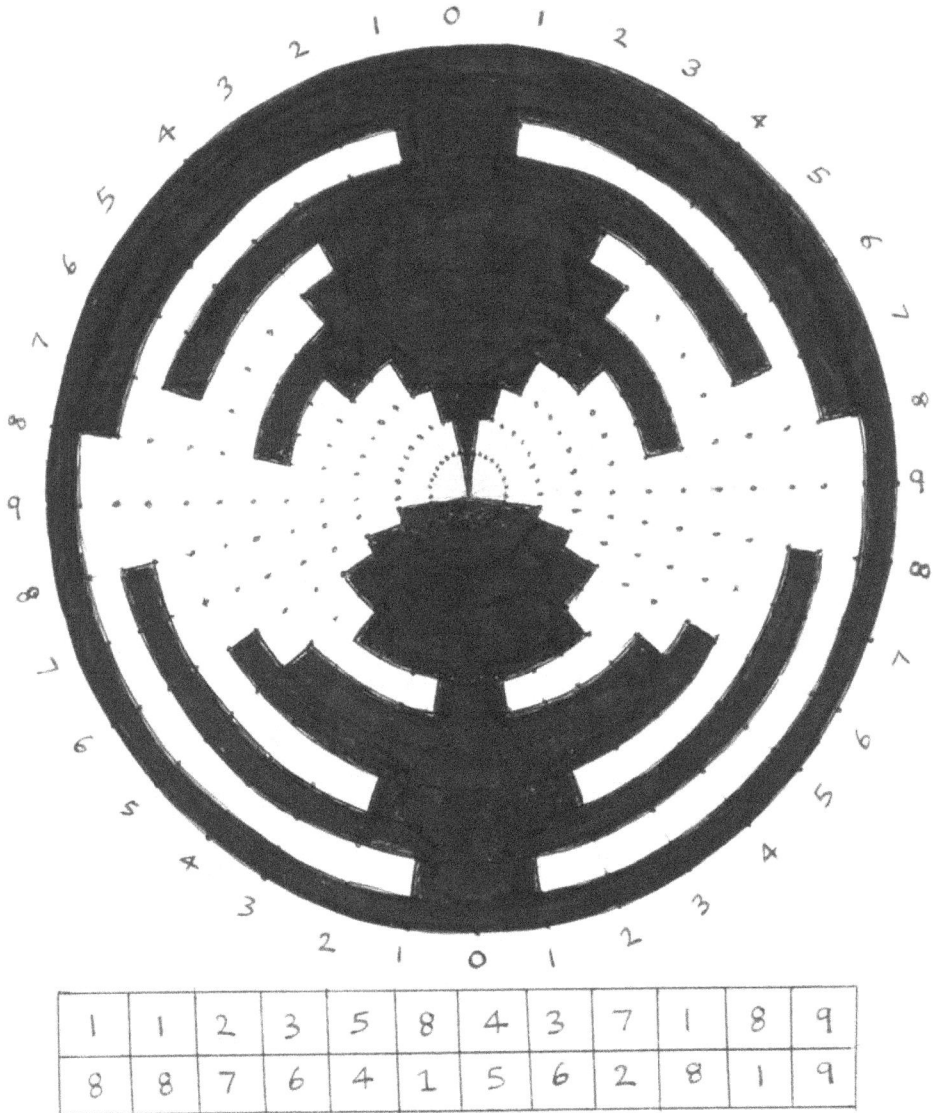

Fig 3e
**Showing the completed construction for the
12 Pairs of 9-ness, using the linear digitally compressed
Fibonacci Sequence Phi Code 1.
The Order of Plotting the 12 Pairs is geared from the Centre
and Outwards.**

Can you visualize what Fig 3e would look like if we decided to reverse the way we plotted these 12 Pairs of 9, by starting with the first pair (1,8) on the outside rim of Fig 3c. As you continue to plot all the pairs, the final pair (9,9) would end up in the heart or centre of the diagram mandala and would have to be filled in or sketched in as a **black hole**! It would also give the overall appearance of a toroidal (ring-shaped) entity, indicating suction to the centre, getting sucked-in! which is exactly the definition of Phi's Force: centripetal implosion.

[**PHI CODE 108 MOVIE / THEATRE**:

Imagine children in the near future learning all of this sacred geometry merely by watching an action packed film, where a crop-circle like Fig 3e appears on the earth, and no-one can decipher what the cryptic geometry means, except for the Magic Square Man. I would actually give this story to a group of school children and let them write the script and act out the scenes theatre-style, with the full intent or purpose of teaching the viewing audience of parents and friends, what is the Golden Mean, how did we compress the Fibonacci Numbers, how did we arrive at the 12 Pairs of 9, and why it is a StarGate portal.]

Do you have a creative or original way in which the 12 Pairs of 9 can be plugged into a circular diagram? as shown in this article designed to inspire your quest for intelligent patterns that can be used practically and for the benefit of society.

When you have blacked in or coloured in Fig 3e, place it 1 meter away from you, at head level, and stare into the mandala, allow any words or feelings to arise and just describe them, no matter how strange it may sound: eg, elegant lines, perfect complementarity, male penetrating female, a couple dancing, humanoid's crown chakra being pierced by an insectoid alien etc write them down and then compare them to what your neighbouring friend conjured.

PART 3

PLOTING PHI CODE 1 PAIRS OUTWARDS TOWARDS CENTRE

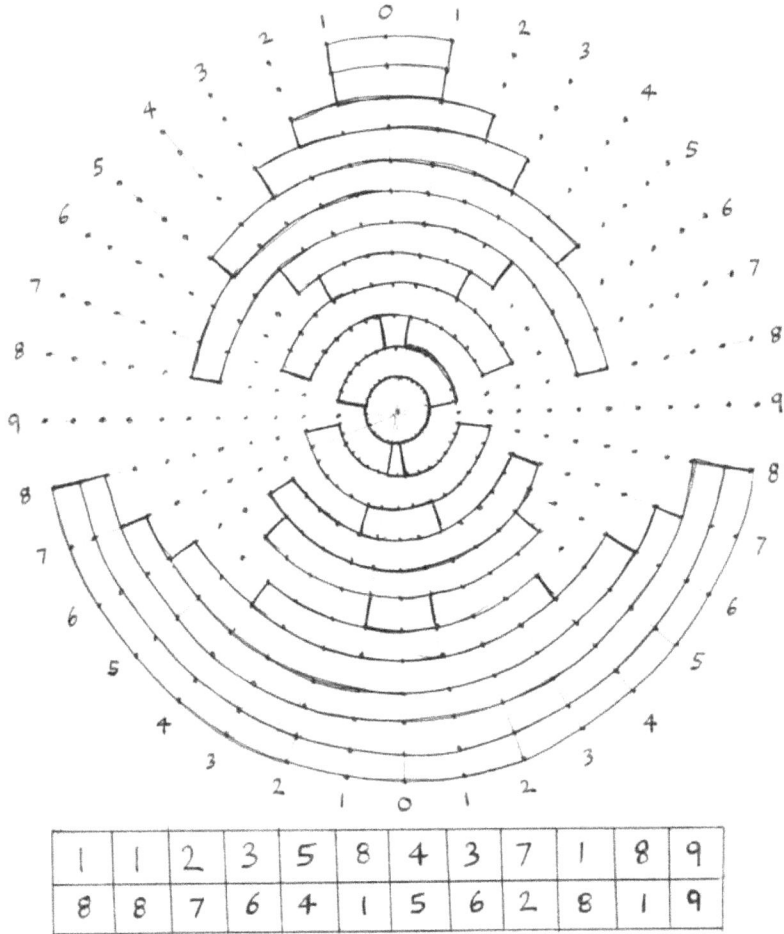

JAIN · MATHEMAGICS WORKSHEET.
GRID/TEMPLATE for PLOTTING the
12 PAIRS of 9ness. (Using 10° radials)

1	1	2	3	5	8	4	3	7	1	8	9
8	8	7	6	4	1	5	6	2	8	1	9

Fig 4a
**Showing the semi-completed construction or outline
for the 12 Pairs of 9-ness, using the linear digitally
compressed Fibonacci Sequence aka Phi Code 1.
The Order of Plotting the 12 Pairs is geared from starting
Outwards then moving to the Centre.**

This diagram and the others that follow, done as an Outline rather than being blackened in, can be used to fotocopy then colour in.

Do you start to get a sense or knowing that the two completed diagrams Fig 3e and Fig 4c evoke different energies or qualities, one is plotting the 12 Pairs by beginning centrifugally from the centre out, then centripetally from the outer rim towards centre. They both have their place in the Phi Code Chronicles. One is not more important than the other. Again, it is another Pair. Do you get the sense that the Phi Code is about Pairings, and has subtle links therefore with our genetic base DNA codings which are based on complementary pairs. What you are seeing here are visual sequences with stop start messages in their codings.

Art Of Jain 1982, "Phi Avatar"

Locus of the Circle, (rolling a circle along the line) generating 12 different sized forms of "Haha-Aha" represented by the Golden Triangles within the Nautilus Shell Spiral.

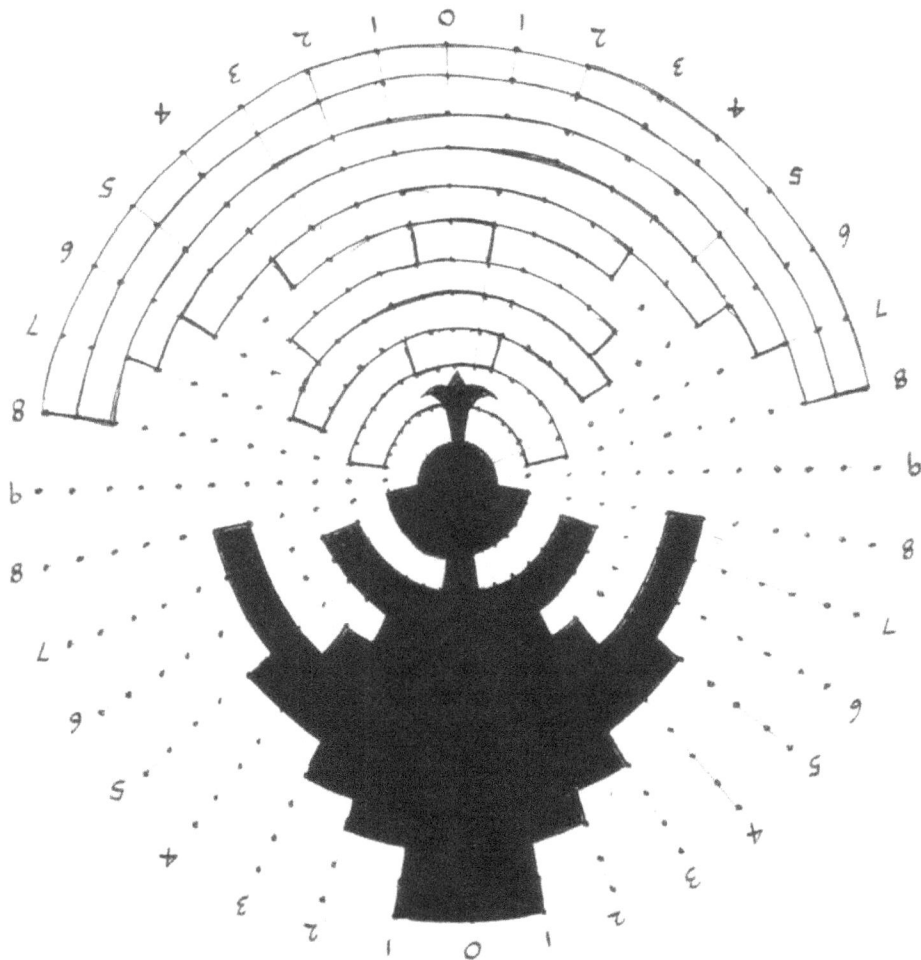

Fig 4b
Showing an Isolation from the above Template
The Order of Plotting the 12 Pairs of Phi Code 1, is geared from starting Outwards then moving to the Centre.
The black hole in the centre is the double 9 Pair (9,9).

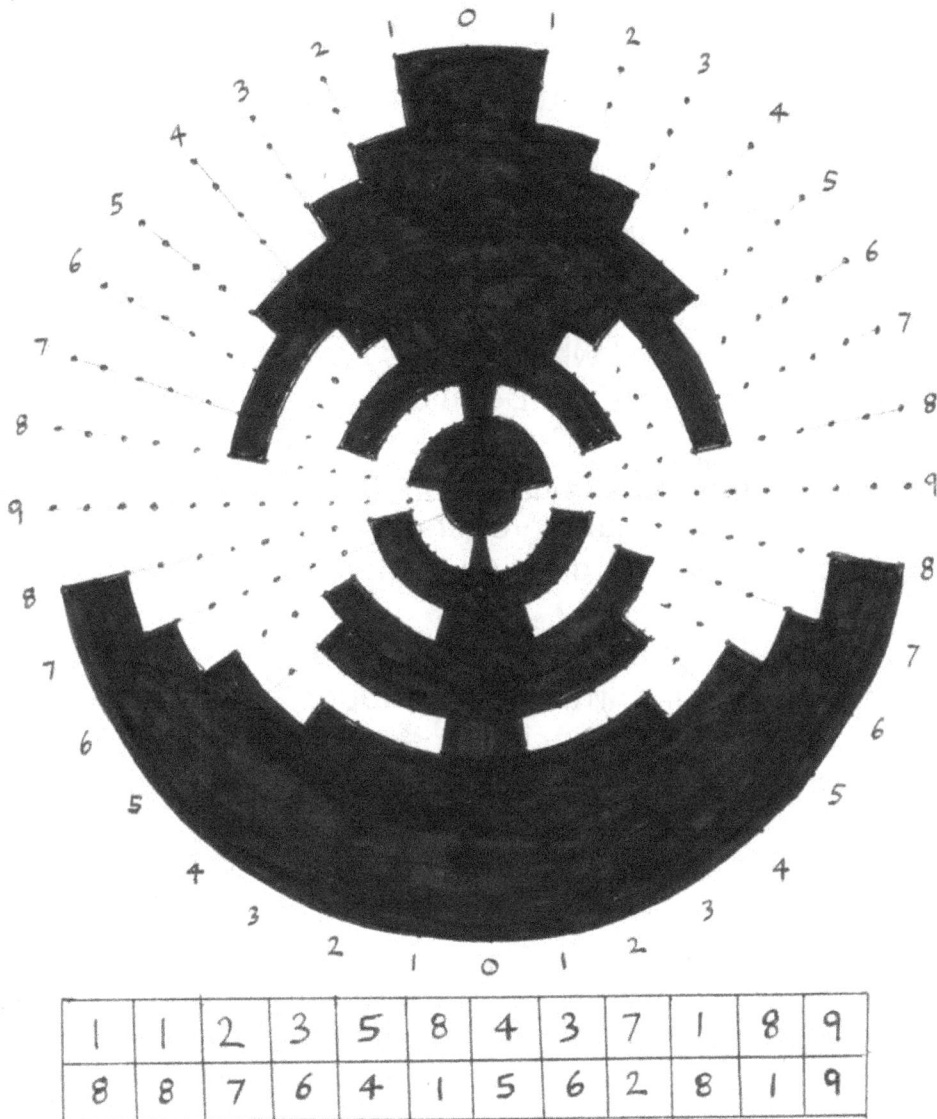

JAIN MATHEMAGICS WORKSHEET.
GRID / TEMPLATE for PLOTTING the
12 PAIRS of 9ness. (Using 10° radials)

1	1	2	3	5	8	4	3	7	1	8	9
8	8	7	6	4	1	5	6	2	8	1	9

Fig 4c
Showing the completed construction or outline
for the 12 Pairs of 9-ness,
using the linear digitally compressed Fibonacci Sequence
aka Phi Code 1. The Order of Plotting the 12 Pairs is geared
from starting Outwards then moving to the Centre.
The black hole in the centre is the double 9 Pair (9,9).

Hold the book in your hand, and turn it around so as to view Fig 4c from different perspectives. It is rich in what it evokes. This is not just any random design, this is based on Timeless Mathematics, what I call **Fixed Design**, Eternal Design. Mathematics can not lie or be fudged, it is a rigid language of the invisible world, making it visible.

It serves as great value to actually mindlessly stare at these images. In ancient India, meditators were instructed by their teachers the art of **Trataka**, to stare at an object or image like a mandala until tears weep from your eyes, then close your eyes and observe the so called "**after-image**". This exercise was designed to awaken or stimulate the Ajna Chakra or **Third Eye** of clairvoyance. Again, observe and record what images come to your inner mind's eye or **Inner Mental Screen**. Definitely, there is a recurrence of the human form, two entities joining towards the centre, there is balance, complementarity, union, angels, children, non-human though celestial beings, women opening to give birth, winged beings, Radha and Krishna dancing a circle dance, and so much more.

I trust that you can recreate these circular Phi Code expressions and release them out into the world. Though somewhat abstract, they still hold this timeless ancient memory of the archetypal order amidst the chaos of our universes, and having them spread around will capture the imagination of the viewers and help to **UNI-PHI** the human species so lost and confused in these troubled times. These ringed phi code templates are intelligent discs imbued now with your own intelligence. Whatever you are thinking of or whatever mood you are in, will be programmed into these psycho-active discs. Thus the real exercise here, whilst drawing these ringed yantra, is to hold focused thought, hold **Pure Intent**. If you program these geometries or blueprints with selfish desires of greed or conquest, you will only amplify, like what the pyramids do, your own destruction. They are to be used wisely for the construction of bridges towards a better world with enlightened connections to Higher Self.

If you have any teenagers in your care, present them with these worksheets and inspire them to recreate these drawings. It is this generation of youth that will take this knowledge globally and bring real change in the modern day school classroom where the buildings are based on nature's designs, where the joy of learning the mathematics that is taught, is an ecstatic process.

The original meaning of education comes from the Latin word "**educe**" which means to draw forth or bring out, elicit, develop; and the Latin word derived from this "**educere**" means to lead forth, bring up.

I pray that these two revealed Phi Codes will accelerate the awakening of our peoples and gear our guiding schools towards the **fibonaccization** of the entire world: One World, One Currency, One Language, One Train Track and of course One Mathematics.

(Rose Carving, showing the definition of Perfect Embedding,
a totally Self-Organized design having another set of 12 Rose Flowers
as its Circumference, symbolic of successfully capable of capturing and sharing
the 12 Pairs of 9-ness inherent in the Reduced Fibonacci Sequence.
Source of Image unknown)

For (y)our reference, it is perhaps of use to actually compare the two completed diagrams Fig 3e and Fig 4c to merely ponder upon what we have just created.

These are disks or yantra or power diagrams chiselled out of the living mathematics of nature.

Fig 3e
Phi Code 1 Plugged into the Circle of 108 Segments
Pairs plotted from the Centre towards the Outer

Fig 4c
Phi Code 1 Plugged into the Circle of 24 divisions
Pairs plotted from the Outer towards the Centre.

SOME CREATIVE TIPS and SUGGESTIONS
for the TEACHER to GIVE to the STUDENT
WHEN DESIGNING the PHI CODE DISCS:

The teacher, supervising their students to draw and colour-in these designs can suggest a few creative options:

Instead of using "black" as I have done to colour-in or shade-in the spaces or forms of Figs 3e and 4c generated, the student can use two colours in replacement of "black & white". The "white" here is the negative space, and its colour can be highlighted.

Fig 3e has a definite "white space" compared to Fig 4c which has no bourndary around its "white space". Try different paired colours to highlight Fig 4c.

Before the drawing class begins, the teacher can show both discs of Figs 3e & 4c to the student, after learning the theory of these designs and are asked to chose to draw 1 of these 2 patterns that they resonate to.

They are also asked to choose any 2 favourite colours to define the "black & white spaces". Towards the end of the lesson, the teacher and students can read or interpret one another's completed discs, eg:

Mary chose Red & Black as her 2 colours and she chose Fig 4c which is the Path from the Outer to the Centre. Like a tarot reading, a student may comment that Catholic Mary is symbolically currently returning home to her Spiritual Centre. She is also still wrestling with passions and lusts symbolized by the Scorpio colours of Red & Black. "Black" here may represent some Bitterness that she is hanging onto, thus Mary's current spiritual challenge or quest is about forgiving someone, including forgiving herself, who acted out some bitterness towards her.

Or Fred, chose to colour in Fig 3e in yellow & orange. The students may comment that he is on a new Journey, which is radiating out from the Centre to the Outer. He is currently studying philosophy and comparative religions with great gusto or excitedness. According to Madame Blavatsky, the colour 'yellow" is the realm of philosophy and high intellectual or mental endeavours. And "orange", like the sun, is high energy, it represents activity and passion etc.

Both Fred & Mary have been "read" by their peers via the discs that they chose. Their yantram (singular of yantra or power art) therefore reflected their current auric landscape, mood or life pathway. It is only symbolic, but none the less, a good exercise to play with. Let the student say anything that arises in their mind to be said. The fact that they have chosen to manifest me or you as a teacher to learn this lost and cryptic knowledge is another important part of the reading, which hints at an evolving higher consciousness and a seeking for the truth.

You may interpret more by looking at:

1 – how neatly did the student draw or construct Fig 3e or Fig 4c? Was it executed with love and care, or rushed and imperfect...

2 – Were the colours chosen "happy" or "sad"?

The disc that you or they have drawn is thus a map of your subconscious world. **Via the Universal Language of Pattern Recognition, we have effectively made the invisible world, of the subconscious mind, visible**!

Art of Jain, 1994
Cosmic Being Guiding the Transmission of the Phi Codes from the Celestial to the Terrestrial dimensions.

PART 4
PLOTTING PHI CODE 1
INTO A CIRCLE OF 108 SEGMENTS

MORE CIRCULAR PHI CODE YANTRA
BASED on the DIVISION of the CIRCLE into 108 SEGMENTS:
9 Concentric Circles and 12 radial divisions (9x12=108)

We have been inspired again, and decided to create another variation of Figs 2abc, the first diagrams known as the Mystic Cogged Wheel where the data of 24 digits were plugged around the Unit Circle, lets say exteriorly or externally or circumferentially.
In this upcoming exercise, using only Phi Code 1, we are going to plug in the 12 Pairs of 9-ness internally into the Unit Circle. Then we are going to examine the differences between each of the 12 Pairs. eg: lets take the first pair in Phi Code 1 which is (1,8). The Difference between the 2 numbers (8 − 1) = 7 so we will shade in that value of 7-ness in the appropriate 1/12th sector or slice of the pie.
Regarding the significant Pair (9,9) the two numbers are subtracted from one another giving (9 − 9) = 0 and will not be shaded in black. The emptiness of this segment creates a lock and key effect for this ancient data to connect and find meaning as like DNA codings.

We therefore need another template (Fig 5a) or blueprint upon which to best capture this intelligent data, and suggest that we divided the Unit Circle into 108 segments being based on 12 Radial lines and 9 Concentric Circles. The 12 Pairs are symbolized by the **DoDecaGon** which has 12 sides, in ancient Greek, "Do" = 2 and "Dec" as in Decimal = 10, thus "Dodeca" = 2 + 10 = 12, and "gon" means "side". The 9 Concentric Circles are based on the Pair values ranging from 1 to 9. It is wise to photocopy this template several times so you can draw the following for yourself.

Jain Mathemagics Worksheet
The CIRCLE DIVIDED
Into 108 SEGMENTS

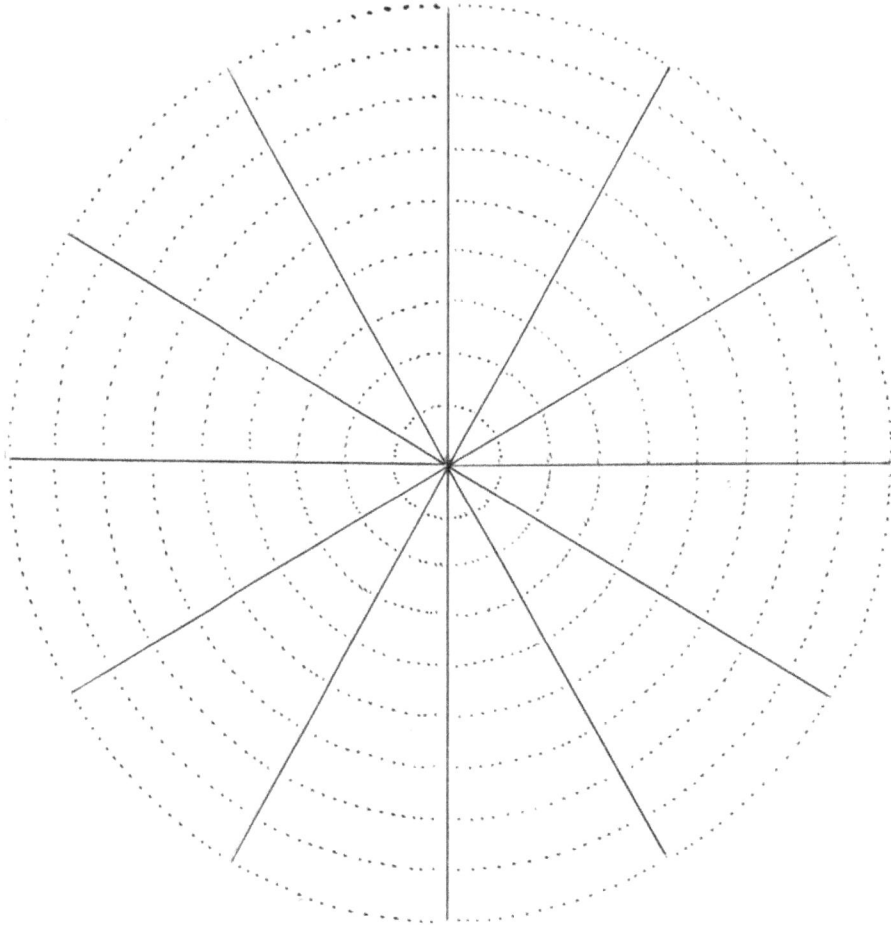

(Composed of 12 RADIALS and
9 CONCENTRIC CIRCLES).

Fig 5a
The Circle Divided into 108 Segments
composed of 12 Radials and 9 Concentric Circles
(typical Worksheet for the Jain Mathemagics Curriculum
for The Global School).

Let us first look at **Phi Code 1** which consists of the following 12 Pairs represented like this:

(1,8) – (1,8) – (2,7) – (3,6) – (5,4) – (1,8) – (4,5) – (3,6) – (7,2) – (1,8) – (8,1) – (9,9).

Now let us list the differences between each Pair:
7-7-5-3-1-7-1-3-5-7-7-0
These differences act like a force between the extremes of the Pair, acting like an electrical **voltage** or a tension that exists in its very nature. Yet, as we will see, there is an exquisite order, a harmony or an innate balance between all of this apparent random-looking chaos of the phi codes.

Before we shade in or blacken the Pair differences, I will first outline or mark in bold in Fig 5b the actual Pairs so we can get an overall view of this cryptic cosmic spiderweb.

Also, to mark these Pairs, we assume that all counting begins from the center point, thus when we mark any pair like (1,8) we count from the centre and mark the first and eighth concentric circles.

You will see in Fig 5b that the 12 Pairs have been written on the outer rim of the 9th circle because that is the specific information that will be plotted in that slice of the pie being a 1/12th segment.

Regarding the position where to start plotting the 12 Pairs on the Circle, the obvious or most conventional starting point is to follow the motions of the clock, thus the first pair in Phi Code 1 being (1,8) will be placed at the 1:00 o'clock position, then continue clockwise.

Also Fig 5b is merely a halfway design, that it, it is not completed until Fig 5c. Sometimes by merely drawing outlines of the arcs, stopping and observing your work at the halfway marks can lead to inspiration or reflection. Its like taking a deep breath, a reflective pause, a moment to inspect what is being created, before it is created; out of the depths of our deep subconscious worlds. It is an emphasis to raise or praise and periscope your head upon the new landmarks, to open more your consciousness and drink in the surrounding environs, as if landing on the moon for the first time; or like a cat playing with it's captured prey, its time to look at the new breed of mice in your net. What have we here? you the detective of patterns. What is the form of this gift of god that I am on the verge of eating, devouring; how pretty are your fine microscopic capillaries running along those shimmering wings says the lizard to the dragonfly in its mouth, yet what I eat or capture I become, for it is I and I am it. Suddenly in this mathematical Oneness, there is no master or slave, no predator or victim. The apparent duality of the Pairs of 9, all these opposing codes were necessary to realize this Unity Consciousness, the ability to **Uni-Phi**.

Jain Mathemagics Worksheet
The CIRCLE DIVIDED
Into 108 SEGMENTS

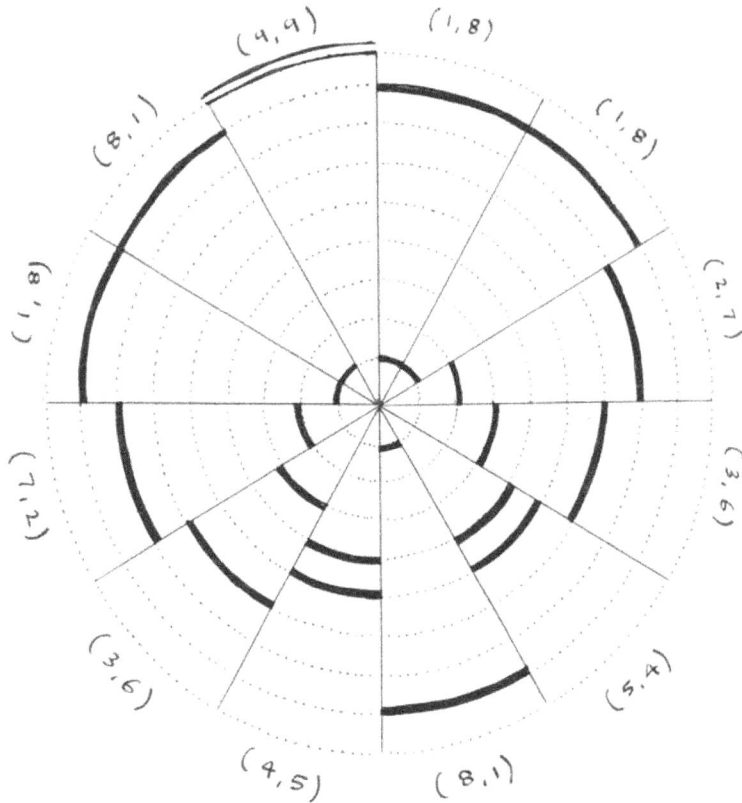

(Composed of 12 RADIALS and 9 CONCENTRIC CIRCLES).

Fig 5b
PHI CODE 1
Plugged internally into the Circle divided into 108 Segments.
The Bold Lines or Arcs represent the numerical difference
between the numbers of the 12 Pairs. (The gaps between
the arcs need to be shaded in to form the completed Fig 5c).

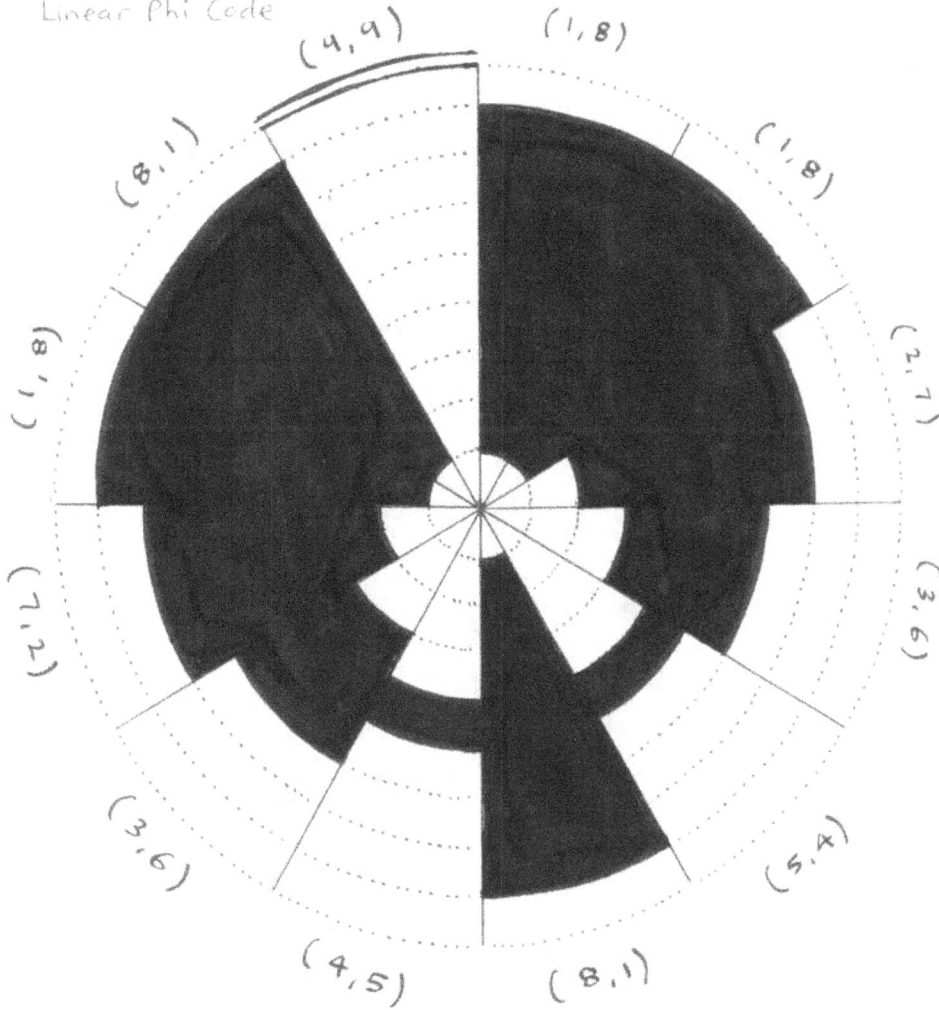

Jain Mathematics Worksheet
The CIRCLE DIVIDED
Into 108 SEGMENTS

Pairs for the
Linear Phi Code

(4,9) (1,8)
(8,1) (1,8)
(1,8) (2,7)
(7,2) (3,6)
(3,6) (5,4)
(4,5) (8,1)

(Composed of 12 RADIALS and
9 CONCENTRIC CIRCLES).

Fig 5c
PHI CODE 1
Plugged internally into the Circle divided into 108 Segments.
The completed shaded areas represent the numerical
difference between the numbers of the 12 Pairs.

PART 5
PLOTTING PHI CODE 1
ONTO THE 9 POINT CIRCLE

To conclude this section, shown in Figs 6a and 6b and 6c,
here is the **Phi Code 1 plotted onto the 9 Point Circle**.
This is the **Translation of Number Into Art**,
the essence of this book.
Question: does this pattern have any symmetry or order?

(Nicholas de Cusa 1401-1464, German cardinal and great mathematician)

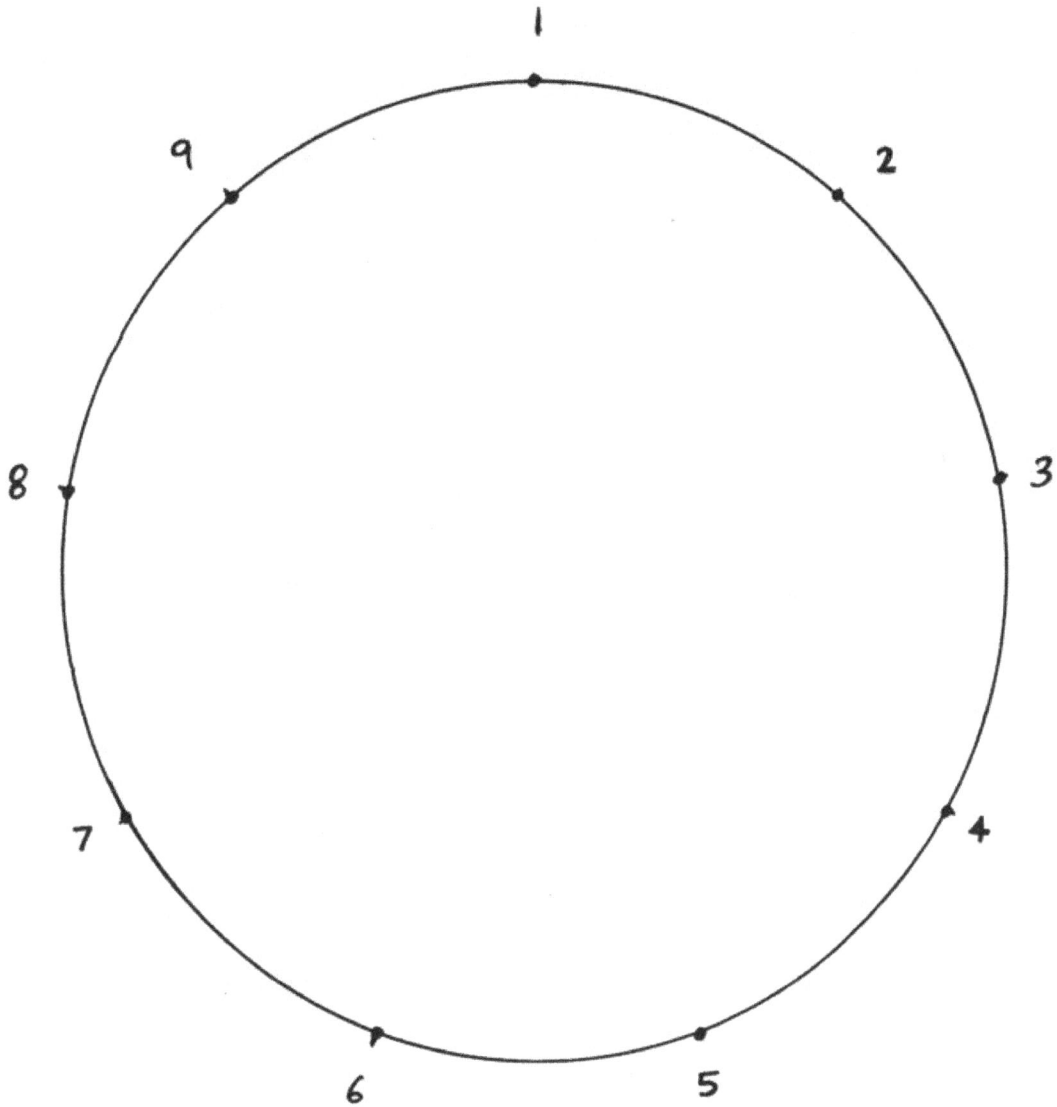

Fig 6a

Plot the Phi Code 1 onto the 9 Point Circle by drawing a long unbroken continuous line from 1 to 1 to 2 to 3 to 5 to 8 etc till the very last number 9; then join this last number back to the first number to close the circuit.

1-1-2-3-5-8-4-3-7-1-8-9-8-8-7-6-4-1-5-6-2-8-1-9

Plotting the 24 Digits of PHI CODE 1 onto the 9 Point Circle. this is the Translation of Number into Art.

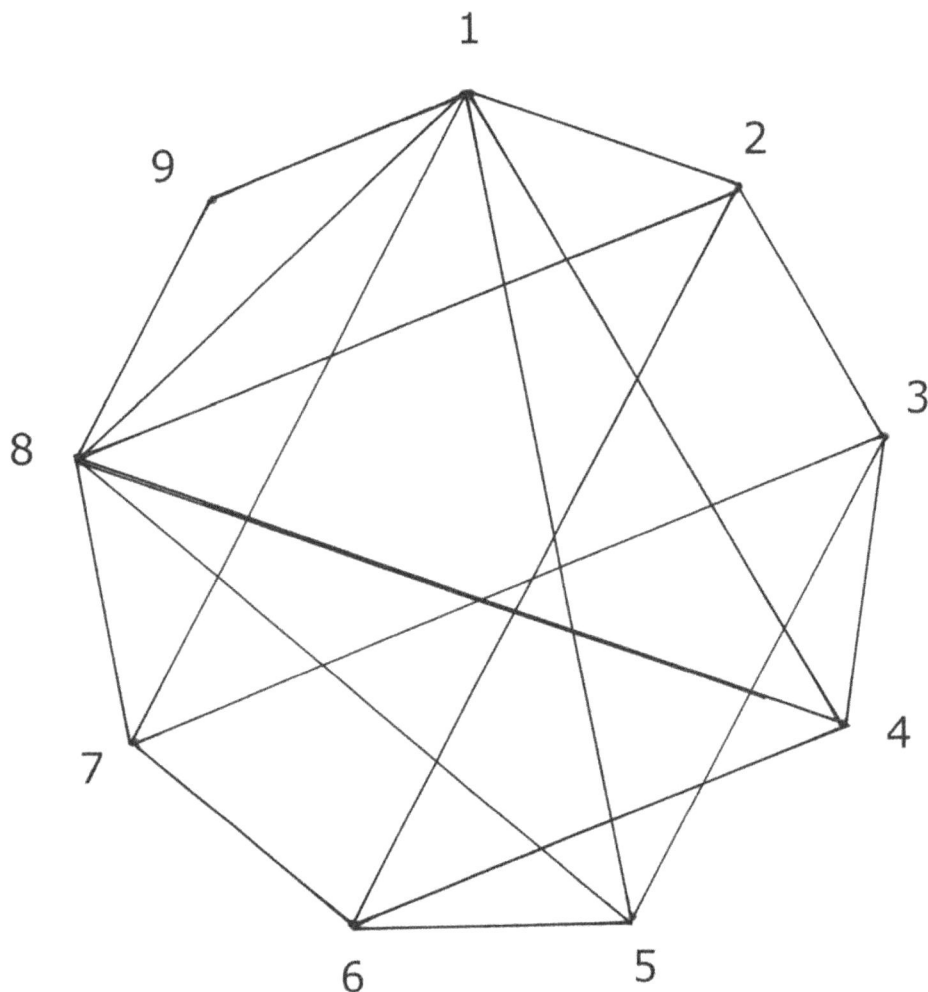

Fig 6b
Solution:
Phi Code 1 plotted onto the 9 Point Circle
(with the number 1 at the north point or zenith)

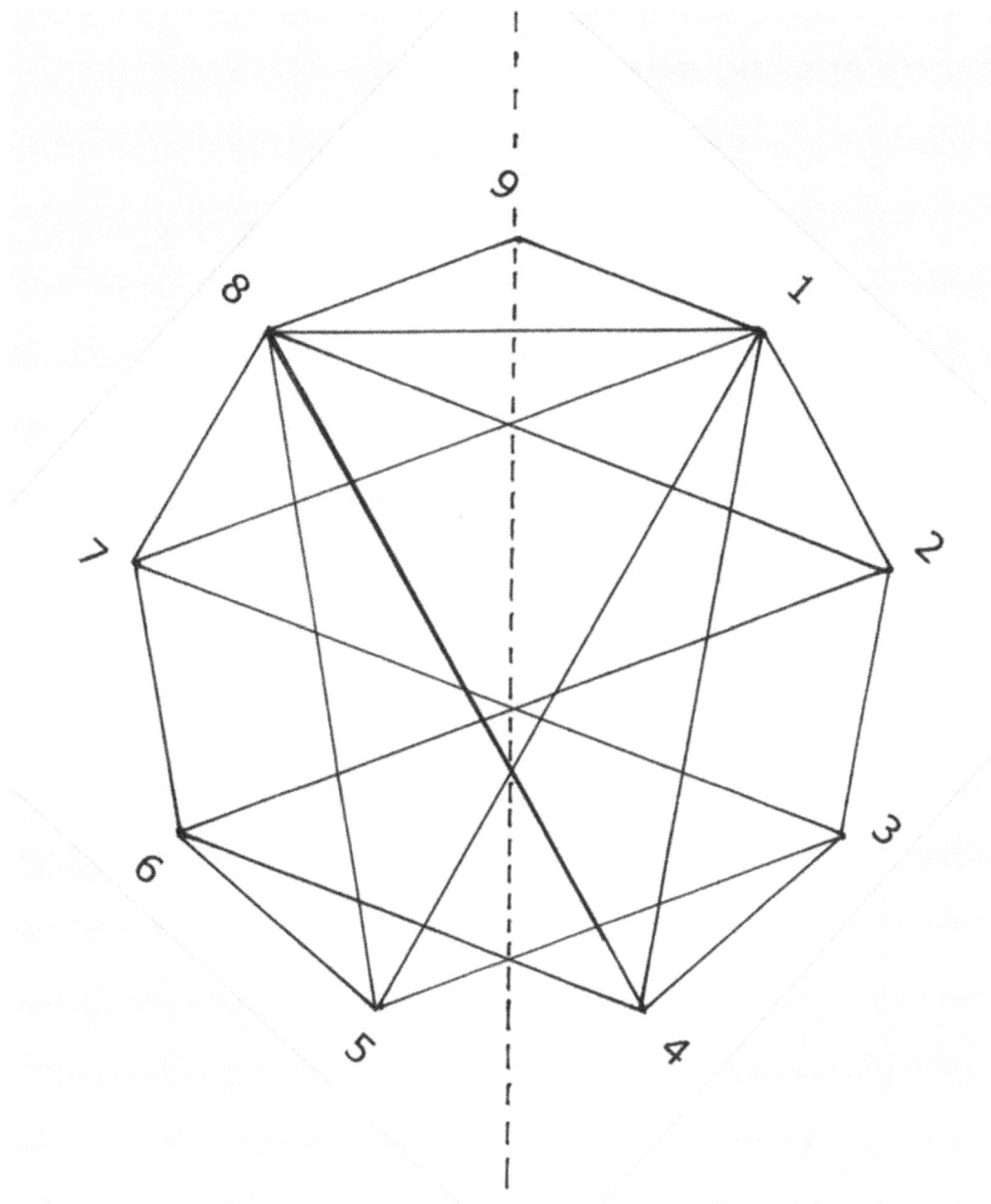

Fig 6c
Phi Code 1 plotted onto the 9 Point Circle
showing a distinct mirror-image symmetry along the dotted
axis running through the number 9 & the midpoint of 4 to 5
expressed as "4.5".
(with the number 9 at the north point or zenith
so that the axis of mirror-imaging is vertically
aligned for aesthetic effect.
This will be compared to Phi Code 2 in the next book for
further comparisons and important realizations.)

THE TWO POSSIBLE PHI CODES 108,

On this page, let us examine:
Phi Code 1 plotted upon the 9 Point Circle and
Phi Code 2 plotted upon the 9 Point Circle
merely for the possibility of looking for similarities or differences.

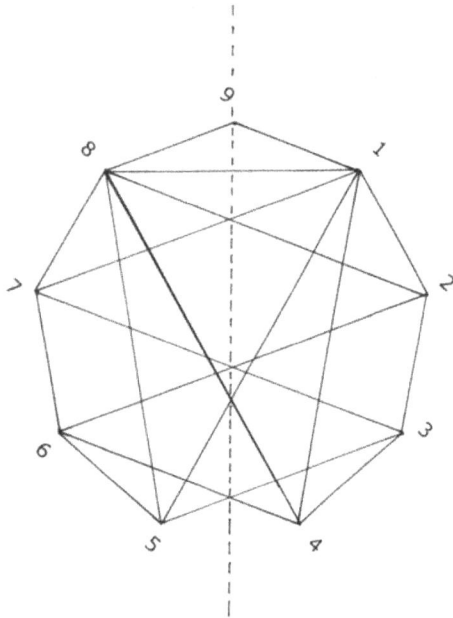

Fig 6c
Phi Code 1 on 9 Point Circle
to be compared to Phi Code 2.

Phi Code 2 on the 9 Point Circle will be revealed in the next book.

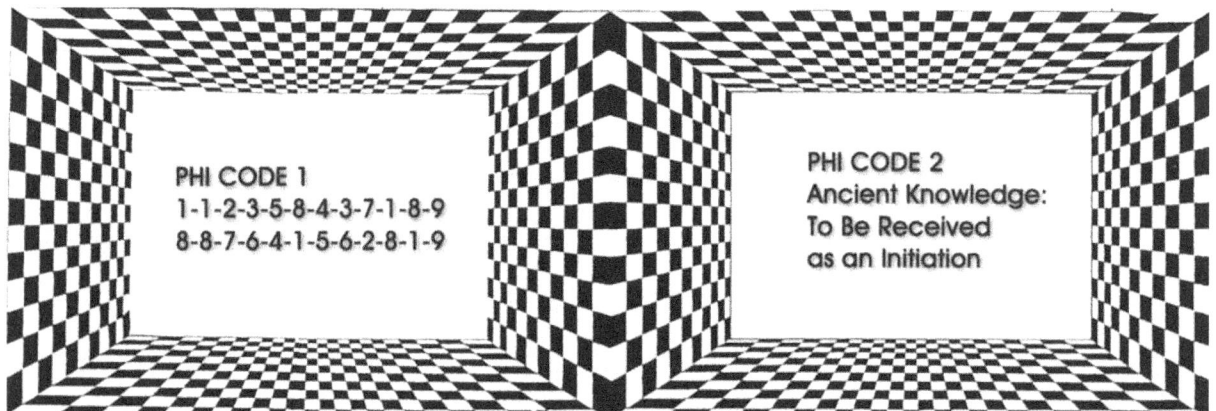

PHI CODE 1
1-1-2-3-5-8-4-3-7-1-8-9
8-8-7-6-4-1-5-6-2-8-1-9

PHI CODE 2
Ancient Knowledge:
To Be Received
as an Initiation

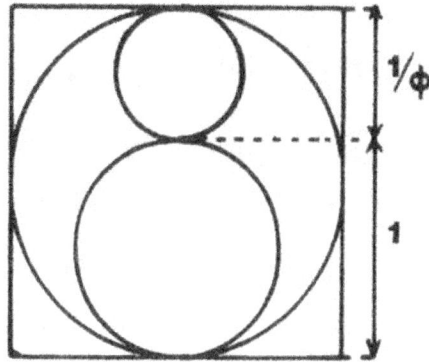

The Arbelos

The "Arbelos" and other "circle geometries" were studied by Archimedes, considered one of the foremost mathematician of all times and the greatest creative genius of the Mediterranean world. (ca 287 - 212 BC)

If AO = 1, AE = Ø
If AB = 1, AO = Ø
If AD = 1, AF = Ø

AO : OE = Ø
AD : DF = Ø
ED : DO = Ø
OE : EF = Ø

Fig 7

The Arbelos, is another excellent example of Mystically Phi Ratioed "Wheels within Wheels".

Of the 3 wheels shown, the internal small one plus the internal medium one divide the largest circumferential wheel into the Divine Phi Proportion!

Addendum:

PHI CODE 2

You have just explored "**PHI CODE 1**"
let's call it linear, as it based on the infinite linear additive sequence
of the Fibonacci Numbers.

Let me put forth here that there is a far more exciting Phi Code,
that also adds up to 108, but its peculiarity or distinction is that it is
a multi-dimensional code.

In this day seminar and booklet on Phi Code 1,
that you have completed and therefore now graduated,
you are able to receive this Phi Code 2 upon your next registration
next season to learn this Knowledge.
I truly believe it to be the most powerful and special mathematical
sequence on the planet, and regard it as an **INITIATION**
even to give it to you.
At this stage, congratulations on having expressed your interest in
the Divine Proportion, and know that there is more to come.
I can only give you a clue now, that this Phi Code 2 relates partially
to the following interesting pattern that expresses the Powers of Phi
in terms of the Fibonacci Numbers!

(first written onto the web by David Thomson in Jul 2006)

$$Phi^1 = Phi$$
$$Phi^2 = Phi(0+1Phi)$$
$$Phi^3 = Phi(1+1Phi)$$
$$Phi^4 = Phi(1+2Phi)$$
$$Phi^5 = Phi(2+3Phi)$$
$$Phi^6 = Phi(3+5Phi)$$
$$Phi^7 = Phi(5+8Phi)$$
$$Phi^8 = Phi(8+13Phi)$$
$$Phi^9 = Phi(13+21Phi)$$
$$Phi^{10} = Phi(21+34Phi)$$
$$Phi^{11} = Phi(34+55Phi)$$
$$Phi^{12} = Phi(55+89Phi)$$

QUESTION: Why is that you state that this Phi Code is to be revealed as INITIATE KNOWLEDGE ?

ANSWER: The reason for the protectiveness around the dissemination of this **Phi Code 2** is that it has never been printed before, nor released to the world,
thus it is like guarded **Celestial Knowledge**,
and as a **Guardian of this Code**,
as a **Farmer of Magic Square Matrices**
my job is to ensure that it goes to those who are **Engrailed**,
who will hold this code
as if they would care for a loved crystal,
and in a similar fashion, program it with **Thoughts of Peace and Love** for the **Fibonaccization of the Planet**,
to bring **Global Unity** above all other costs and factors.
Your **Heart is Pure**, your **Intent is Pure**,
thus it has come to You.
How you work with this Phi Code 2
is your personal challenge, whether you attribute the numbers to specific frequencies or musical notes,
or build it as a children's playground
or make it into a 108 beaded rosary
or use it in your building constructions
or colour code as tiles on your floor,
it is up to you.
This code is strengthened by your **buddha field**,
your love and intent.
One condition for receiving this
is that you do not pass it on to others ad lib, as I would prefer that any student wanting this Higher Code needs to jump through some hoops or loops, referring to them to attend
the **12 day Jain Mathemagics Seminars or Immersives**
so they **grok** the pure principles of
Translating Numbers Into Art
The **Universal Language of Shape and Pattern**
The Joy and innate **Beauty of Numbers**
The inter-connectedness of all life
to **SEE GOD IN ALL PEOPLE AND ALL THINGS**.

(JAIN 108   December 2008, Mullumbimby, Australia)

(One of Jain's early phi drawings, whilst living with Australian Aboriginal People in the Torres Strait Islands, 1982. I learnt more about sacred geometry those days, by associating with medicine men, who could not talk a word of English, just transmitted it to by showing me the roots of trees, barks, leaves, medicines etc).

Art Of Jain 1981

"Haha-Aha " Avatar appearing in the Phi Rectangle

Becoming Phi Imbued.

This is what your Living Field looks like

when you receive Phi Code 2

APPENDIX

INCLUDES:

PART 1 – **How to Find the Phi Ratio in the 3-4-5 Triangle.**

PART 2 – **Bibliography**

PART 3 – **Chapter Contents for The Book of Phi,**
Volumes 3, 4, 5 & 6
– **Some Early Art by Jain using Phi Spiral**

PART 4 – **Some Early Rejected Phi Drawings.**

PART 5 – **Pentacle Crack at Base of Glass Jar**

PART 6 – **PHI CODE and VEXILLOLOGY**
(study of the Heraldic Hierarchy)

PART 7 – **Root Phi Conic Sonic Tune Up**

PART 8 – **Why 24 ?**

PART 9 – **Phi to 20,000 decimals**

HOW to FIND the PHI RATIO in the 3-4-5 TRIANGLE

By simple compass and ruler construction upon a 3-4-5 triangle, we arrive at specific ratios of lengths equal to the golden phi ratio and its square, and including the ratio of consecutive Fibonacci numbers. This geometry is a true mathematical gem.

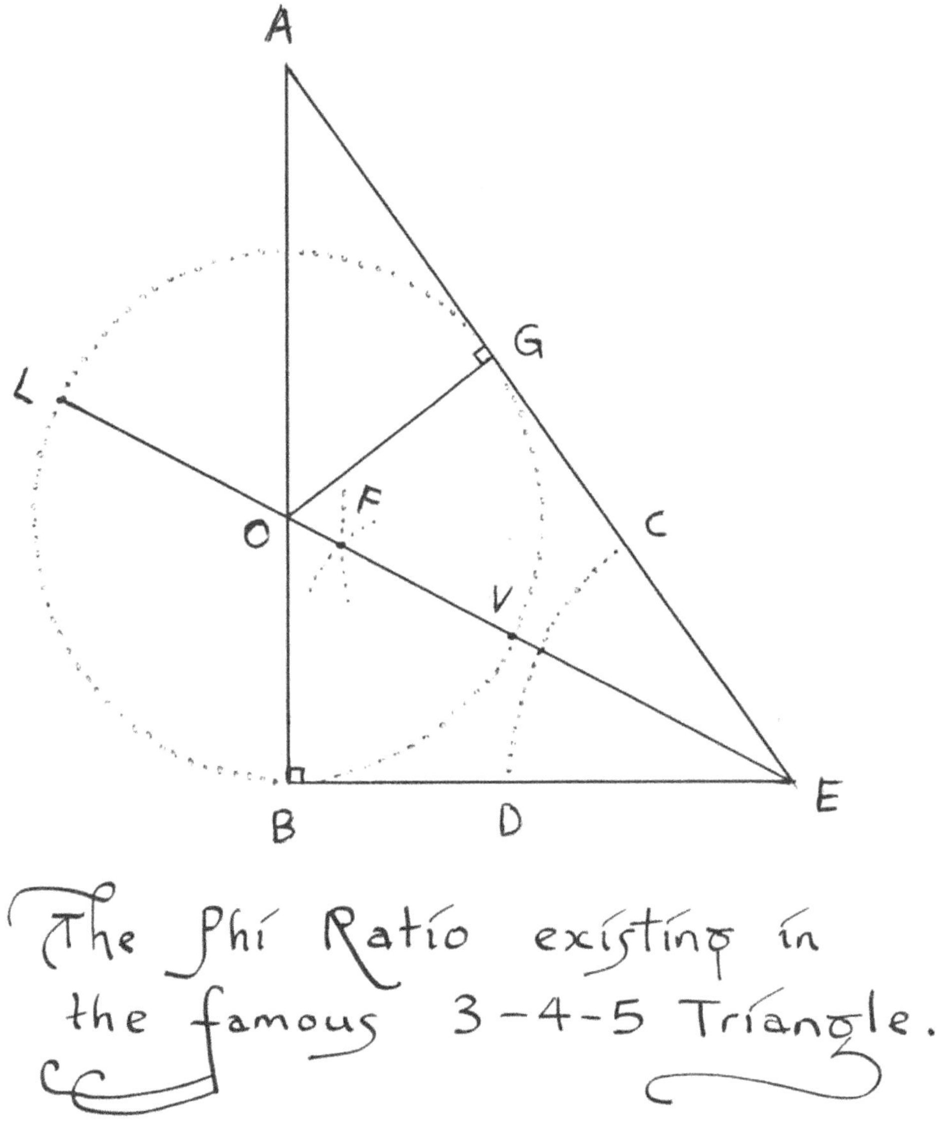

The Phi Ratio existing in the famous 3-4-5 Triangle.

Fig 19

Phi hidden in the 3-4-5 Triangle, also found is Phi squared, and Fibonacci Numbers ratios. Also Root 5, but not shown. (Diagram originally by H.E. Huntley and then Lorena Loo)

Method Of Geometric Proof:

Construct ABE as the 3-4-5 Triangle, with measurements like BE=90mm, AB=120mm and AE=150mm, where AB is perpendicular to BE. First step is to bisect Angle AEB.

(Easy to do. To bisect, draw any arc DC of any size, preferrably like 50mm ie: DE=CE= 50mm. Draw two arcs from these endpoints at D and C by placing the same distance of arc on C, and making another arc from D with the same 50mm arc, and F is the intersection point. Draw EF which bisects the Angle AEB).

Extrapolate the line FE to the left hand side so that it touches AB. "O" is the point where it hits the perpendicular AB. Place compass at point "O" and make a circle with radius r=OB. Extrapolate OE again to the far left hand side till it hits the circle at L. Observe the straight line LOVE. From "O", draw a short line that is perpendicular to the hypotenuse AE. The perpendicular hits AE at G. That is the end of the basic construction.

Here are the following Phi and Fibonacci Number relationships:

EL÷LV = Phi (this reads as the distances EL divided by LV = Phi or 1.618033...)

LV÷EV = Phi

EL÷EV = Phi squared = 2.618...

BE÷BO = 2÷1 = 2

EG÷AG = 3÷2

AE÷EG = 5÷3 = AB÷BO

AB÷AO = 8÷5

LV = 3

EL = 3Phi

One reason why I have just shown you where Phi exists in the famous 3-4-5 Triangle, is that in my next works, I would like to show you how Phi exists in the **Binary Code** of 1-2-4-8-16-32-64 etc and also show you how Phi exists in the **Equilateral Triangle** and therefore in the **Star of David**, and also how Phi exists in the world of Magic Squares. We have just shown in this chapter how Phi exists in the Prime Numbers, so basically the point is that the omnipotent and omnipresent **Phi** is ubiquitously found in all important geometries. This theme would require another booklet of data, so it is not presented here, just only need to know that it all can be mathematically proven.

(Fig 19 taken from and slightly modified from Lorena Loo's website: http://sacredcirclecosmos.com/Sacred%20Geometry/ropestrechers.html

The mathematician H.E. Huntley briefly touches upon this construction and the relations as well in his book *The Divine Proportion*. But she claims it has several errors. She states also that the root 5 ratio is also within this diagram but not shown here).

APPENDIX PART 2:

BIBLIOGRAPHY

- "**Geometry of the Golden Section**" by Robert Vincent, 1999
- "**The Science of Christian Economy**" by Lyndon LaRouche, Jr. 1991
- "**The Order of St John**" (A Short History) by E.D. Renwick, first edition 1958. Subtitled: "The Grand Priory in the British Realm of the Most Venerable Order of the Hospital of St John of Jerusalem". (see chapter on Prime Numbers)
- "**Magic Square and Cubes**" by W.S. Andrews , 1917, Penguin
- "**The Mathematics Of The Cosmic Mind**" by L. Gordon Plummer; 1970, Theosophical Publishing House
- "The Hidden Life In Freemasonry" by Charles W. Leadbeater's 33° book Theosophical Publishing House ,1926 (image on page 31)
- **Aya's 108 Starwheels**: http://www.starwheels.com The new Online School of Sacred Geometry. www.schoolofsacredgeometry.org
- **Srimad Bhagavatam** (image on page 38)
- http://demonstrations.wolfram.com/PatternsFromMathRulesUsingComplexNumbers/ (image on page 3)
- **Dan Winter** www.goldenmean.info
- "**Baalbek**" by Friedrich Ragette (image on page 56)
- **Golden Number** website: www.goldennumber.net (table of "Phi to 20,000 decimal places" on page 186 Appendix 8)
- "**Patten and Spedicato**" research on the historical references to the number 108
- "The **Penguin Dictionary of Curious and Interesting Numbers**" by Wells, D.   London: Penguin Group. (1987): 134
- [Credits for image on p93: The **Ulam Rose** of 1 => 262,144 used here is an embellishment of an image originally created by Jean-François Colonna ©1996, CNET and the **École Polytechnique**, Paris France.

  The picture used here comes out of *Cracking the Bible Code* by Jeffrey Satinover, M.D.]
- Kathleen McGowan: "**The Expected One**", "**The Book Of Love**" and "**The Poet Prince**". Published by Simon and Schuster 2010, New York.
- "**Numbers Of Light**" by Jason O'Hara, 2007 (image on page 109)
- "Yantra" by Madha Khanna, pub. 1979 (image on page 47)
- "The Secret Teachings Of All Ages" by Manly P Hall (reference on p186)
- The Pistis Sophia" by JJ Hurtak. (reference on p186)
-

APPENDIX PART 3:

CHAPTER CONTENTS FOR THE BOOKS OF PHI VOLUMES 3, 4, 5 & 6

Jain's Phi-Pi Connection:
The True Value of Pi = 3.144...

"Lady Nada" by Lily Moses.

**... receding
of the personality
into nothingness**

- LADY NADA

THE ORIGINAL WORKSHEET THAT REVEALED THE PHI CODE

Here is a sample page or worksheet taken from my first book of phi, and used in a vedic maths book (called The Magic Of Nine) to illustrate the Sutra known as Digit Sums or what I call Digital Compression, where the students discover for themselves the 12 Pairs of 9. It was used originally as an element of surprise or Mathematical Discovery, rather than telling the student straight away that there exists a 24 Repeating Pattern that can be viewed as 12 Pairs of 9, whose basic pattern is "108-9" that repeats forever.

Learning via Discovery is the ultimate fashion in how to teach a child the Joy of Numbers.

1·6180339887

the FIBONACCI SEQUENCE	EXERCISE 3.7

Here are the first 24 numbers of the Fibonacci Series, the Living Maths of Nature. The Phi Ratio 1:1·618 is obtained by dividing terms like: 610 ÷ 377.

1, 1, 2, 3, 5, 8, 13, 21, 34, 55, 89, 144, 233, 377, 610,

987, 1597, 2584, 4181, 6765, 10946, 17711,

28657 and 46368. Find the D.S.S.P. or Digital Sum Series Pattern by writing the answer in the box below. The Periodicity = 24.

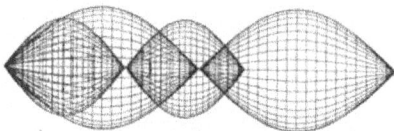

The Fibonacci Series is part of our DNA Genetic Encoding.

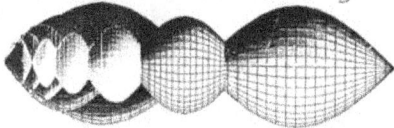

_ _ _ _ _ _ _ _ _ _ _ _

_ _ _ _ _ _ _ _ _ _ _ _

Look very closely at these 2 lines of Digit Sum numbers, and what do you notice?

Photograph of a Pine Tree in Yamba, by Jain,

embodying the exact form of the Phi Triangle!

(notice the "rainbow burst" effect that is not photoshopped,
but taken directly with my digital camera, using my own energy field
to amplify the invisible energies around the tree.
Dr Harry Oldfield, developer of the Kirlian Photograph, of the UK
gets the same images using his advanced PIP camera that can see
clairvoyantly chakras, meridians, phantom limbs, ghost entities etc).

SOME EARLY REJECTED PHI DRAWINGS

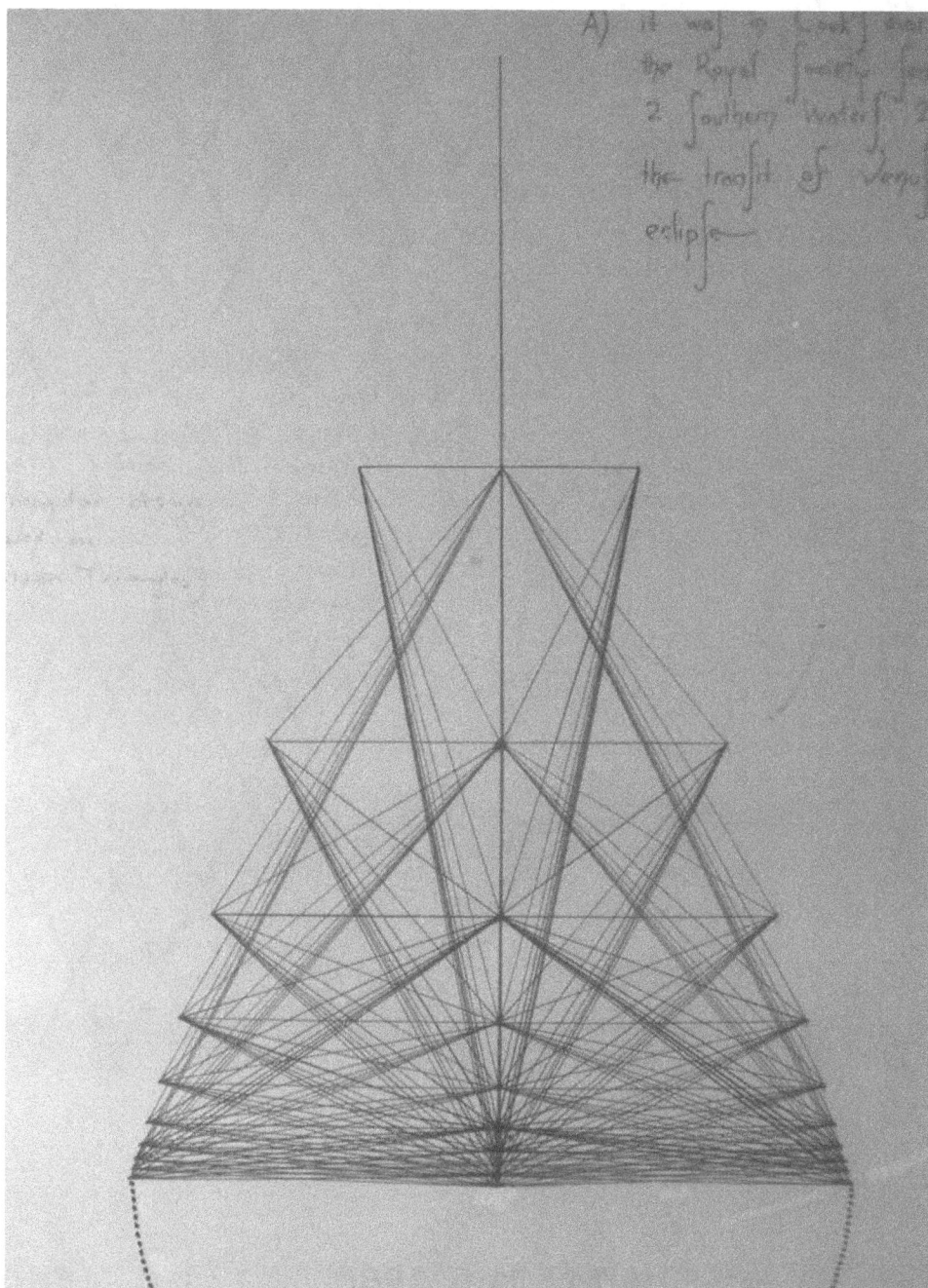

Art of Jain, 1982, "Ship At Sea"
based purely on The Fibonacci Numbers
along the central vertical mast.

Art of Jain, 1986

"Mother and Child" expressed as Phi Spirals

Art of Jain, 1986,

"Phi Spiral Family Entities"

PENTACLE CRACK AT BASE OF GLASS JAR

How the Pentacle 5 pointed Star, which is the masculine form of the feminine Phi Spiral, is also the emblem of 1.618033... proving this by showing an example directly from Nature, how Nature chooses this pentacle phi form above all others. Here are 3 photos to describe what I witnessed.

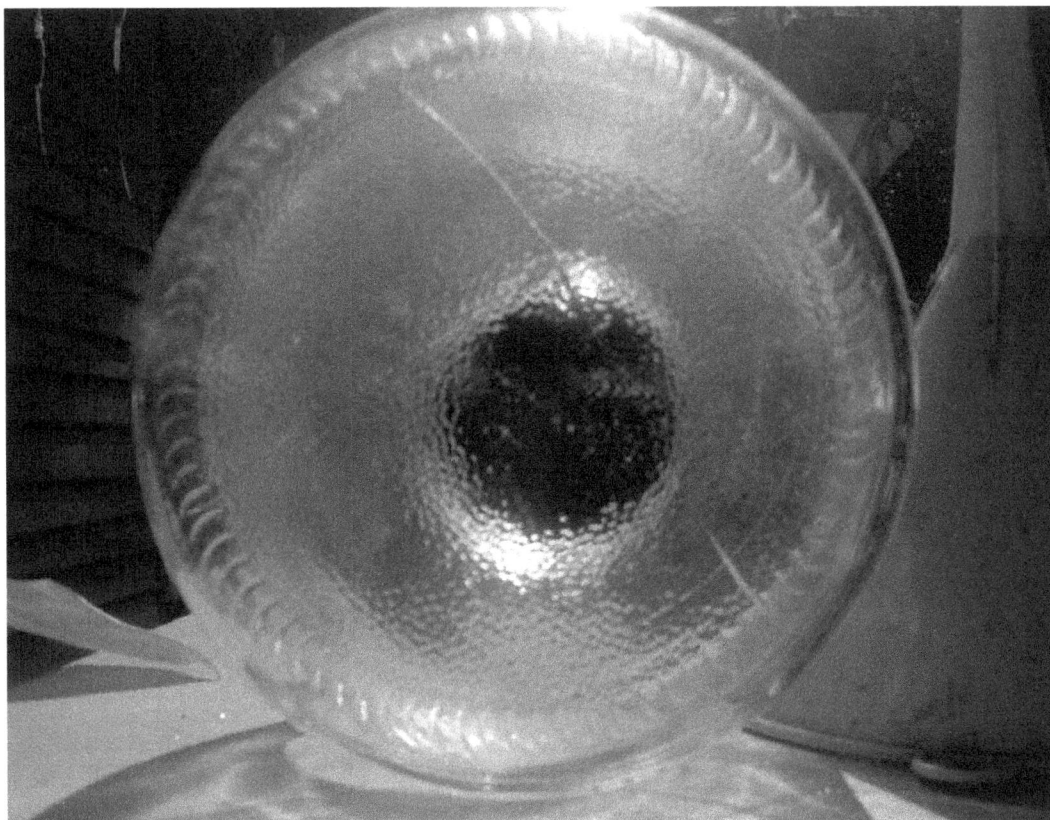

Photo 1 of 3.

This is the base of a large glass jar, having a base of nearly 5.5 inches or 140mm. I dropped it and if you look closely you will see that the crack forms an 5 pointed star! (I have seen this happen in other shock wave incidents. Eg: as a builder by trader, actually a brick-layer, I was doing some carpentry, and nailed a nail through a painted wooden shelf. The paint cracked in exactly the same way as above.

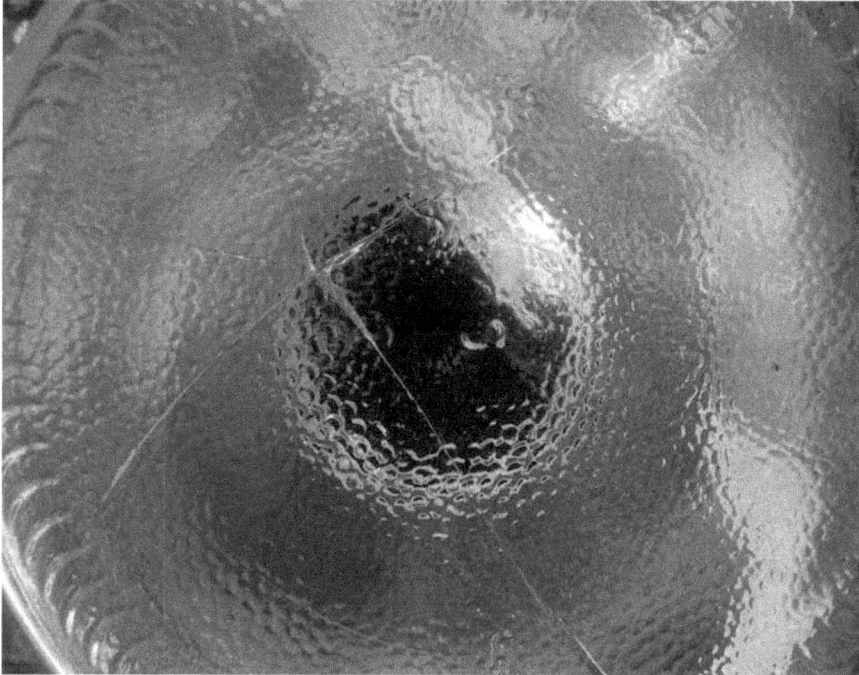

Photo 2 of 3.

A closer view of the pentacle glass crack

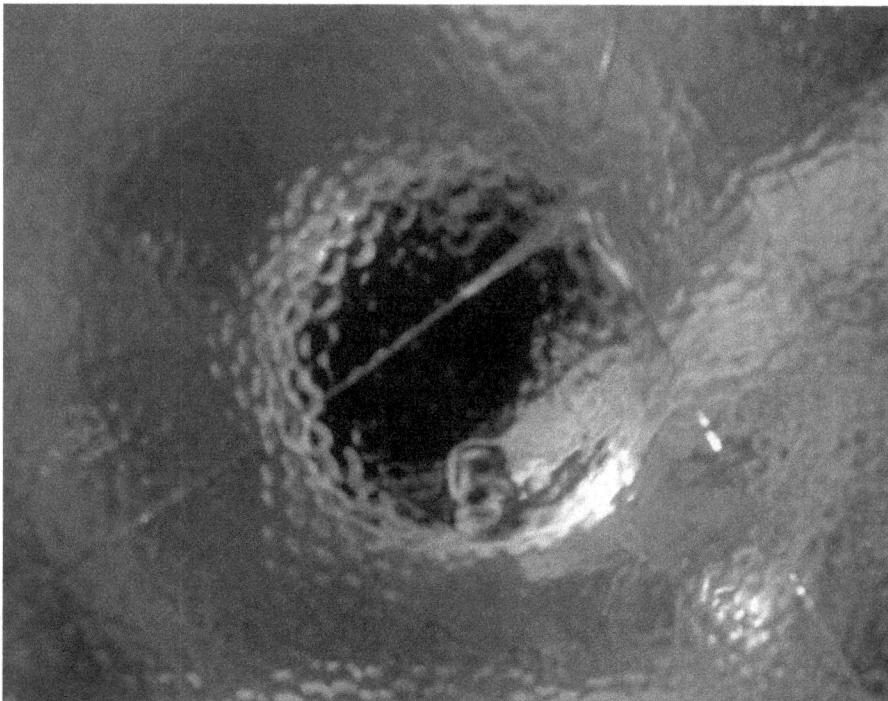

Photo 3 of 3.

Another view of the **pent crack**. It is time to **RE-PENT.**

PHI CODE and VEXILLOLOGY
(study of the Heraldic Hierarchy)

The time is ready now for the creation of a new Heraldic Shield or Banner or Seal or Crest, a way to globally acknowledge the ushering in of the ancient 108 Phi Codes in contemporary times,

Perhaps it is up to YOU to bring forth this 108 Shield displaying both 108 Codes, singularly or as combined wheels or disks.

A SHORT DESCRIPTION and HISTORY of VEXILLOLOGY

◇ Heraldry symbols from Medieval times etc, like the Fleur de Lily, on display, strengthened family ties through marriage. It helped identify royalty, counts, dukes, duchesses, landowners, knightly ranks and "femme and baron" (meaning wife and husband).

◇ In fact, early graffiti showed Crusader's family symbols carved in stone in Bethlehem.

◇ Heraldry also strengthened brotherhood alliances. Armorial garments used to proclaim a person's noble status and these arms were inherited by future generations.

◇ Shields or Family Arms also denoted family member's places in large wealthy or royal family units.

◇ Many wealthy merchants sought these "trappings of gentility" to have carved in stone as their symbol. But they had to pass the rigid tests of the heraldic authorities.

◇ Often a family crest, say for a clan chief, bore space for their motto or words defining their heraldic tradition.

◇ A flag, like the one shown below, or shield determined status, family origin and belief systems. At a time when armies were fighting, and faces were hidden behind armour, these heraldic crests (a 3-dimensional symbol on a helmet) indicated a person's position. Later these symbols became indicators of cultural and national identities, of militaries, clubs, churches, guilds, clubs and governments.

**Isabeau of Bavaria arriving in Paris
for her marriage to Charles V1 of France.**
You can see the large impaled arms
of heraldic shields being displayed.

JAIN 108 SEMINAR
ROOT PHI CONIC SONIC TUNE UP

This is to be performed at the beginning and end of a Jain Seminar. Any two sounds can be used, but these two given are the most appropriate in terms of Short and Long Wavelengths that embrace The Above (Dddoo sound) and The Below (Ahhh sound) respectively.

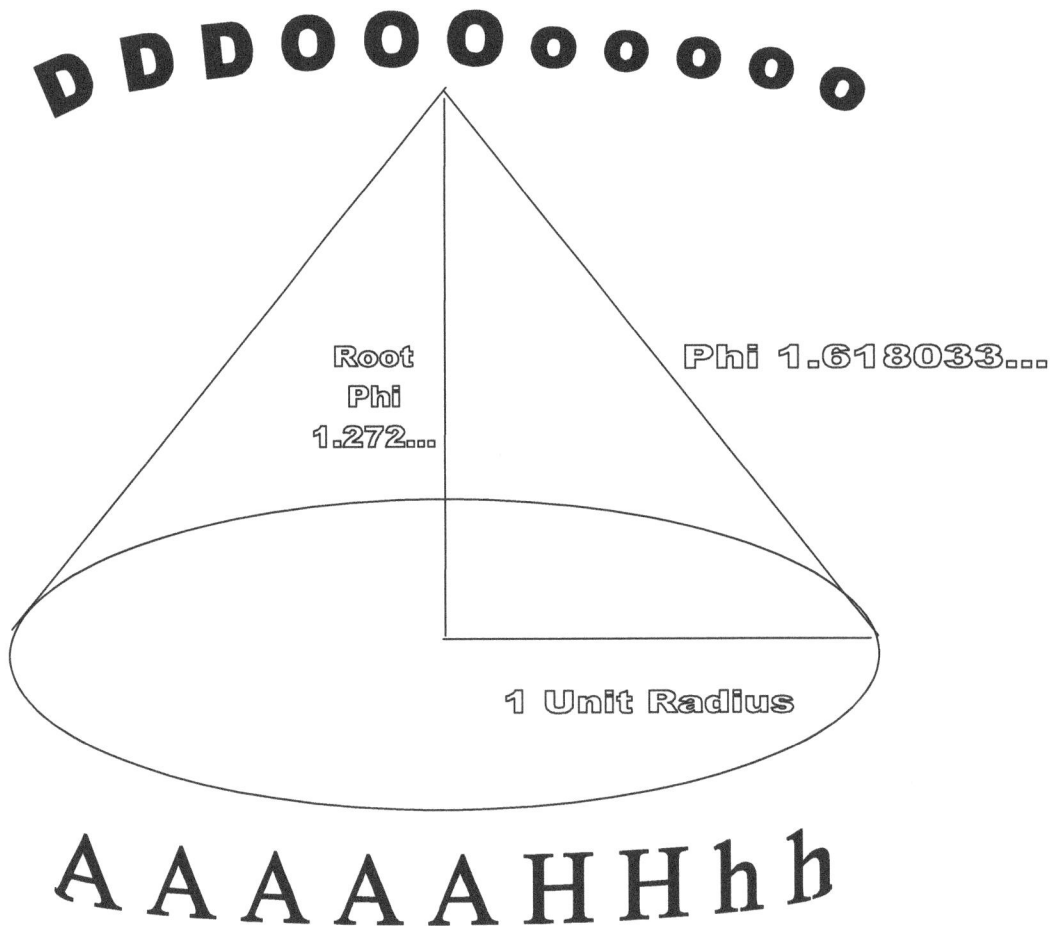

DDDOOOOooooo

Root
Phi
1.272...

Phi 1.618033...

1 Unit Radius

AAAAAHHHhh

Fig 1
Jain 108 Sonic Root Phi Conic that embraces Dualities to reach a state of Oneness. This is the great Etheric Structure that surrounds you to assist the practitioner into entering Sacred Space.

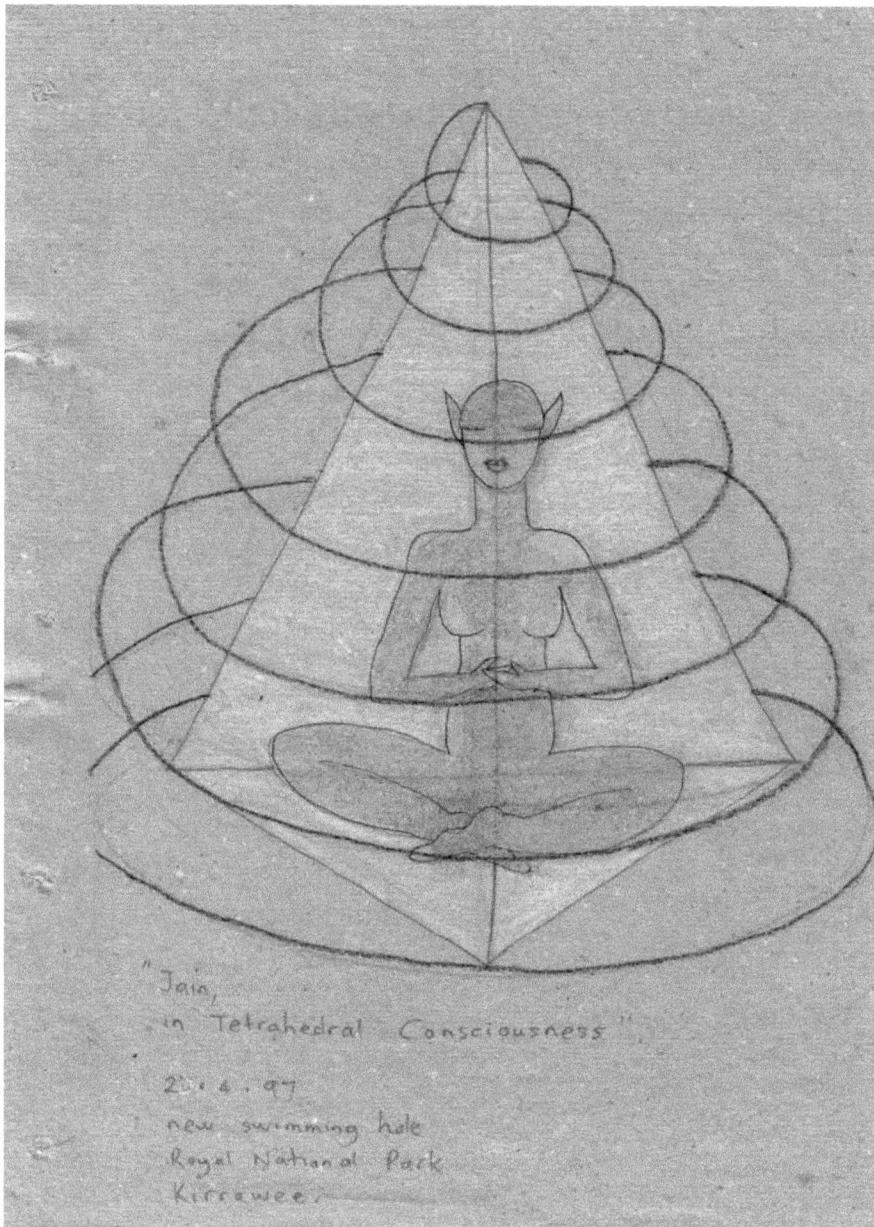

In the handwritten caption within the drawing:

"Jain,
in Tetrahedral Consciousness"
20.4.97
new swimming hole
Royal National Park
Kirrawee/

Fig 2
An original drawing by Jain 1997, showing a meditation accessing the Tetrahedral Consciousness.
Fig 1 therefore depicts the shift from Tetrahedral to Phi Consciousness, specifically downloading the Harmonics of the Root Phi Right Angled Triangle in Fig 1,
the most famous or anointed of all Triangles.

Visualize that you are standing or sitting at the base centre or circle centre of this Phi Cone identical to the inside Measurements or Harmonics of the Gizeh Pyramid.

Visualize that this shape is a room-sized Etheric Structure around you.

You don't have to be a Singer to evoke these two specific tones.
Just get into your own primitive internal space or creature, whether it be an animal, a fusion like a centaur, a bird, insect or primal primate Neanderthal.

As you chant Mindlessly, visualize that the Spirals of Energy traveling from around the base of the cone towards the Apex of the Cone.

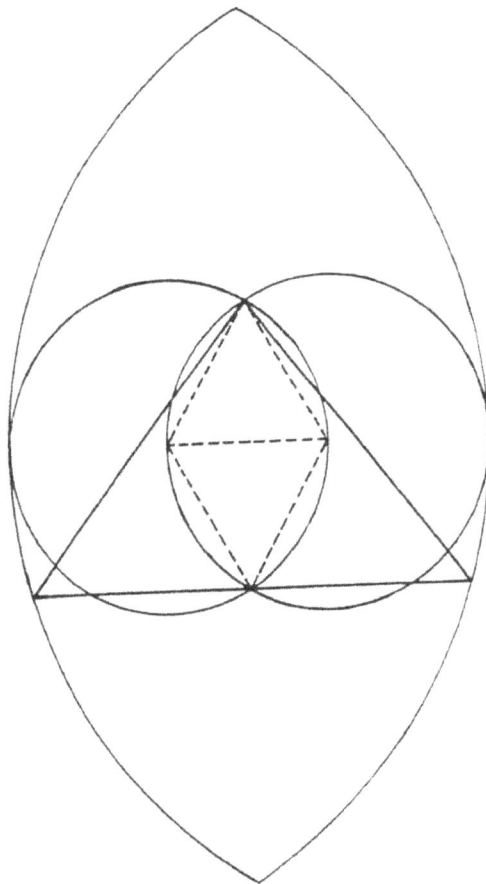

1. The Vesica Piscis with the outline of the Great Pyramid.

JAIN 108's SONIC CONIC For MULTI-DIMENSIONAL PURIFICATION & ACCESS	
LOWER CIRCLE of AAHH	**APEX POINT of DDDOoo**
Start with AAAhhh	End with DDDOoo
Tuning Down	Tuning Up
Accessing	Releasing
Collecting All Ancient Memories from your Pool	Letting Go of All Dross from your Cone Point
Primal Grunting	High Frequency Sounds
Long Wave	Short Wave
Raw and Wild	Refined and Focussed
Vulnerable Feelings	Invulnerable Feelings
Emotional Database	Non-Emotional Expression
Gathering All Attached Entities from All Lives and Dimensions	Exorcising All Attached Entities to Claim your Sovereignty
Looking at All Possible Fears In the Face	Fearless
Cry if you need to, an opportunity to Learn How to Cry	Laugh
Awareness of Dualities Mind	Non-Dual Oneness No-Mind
Keep adding your extra notes here…………........................	

Beauty

Wisdom

Intelligence

Harmony

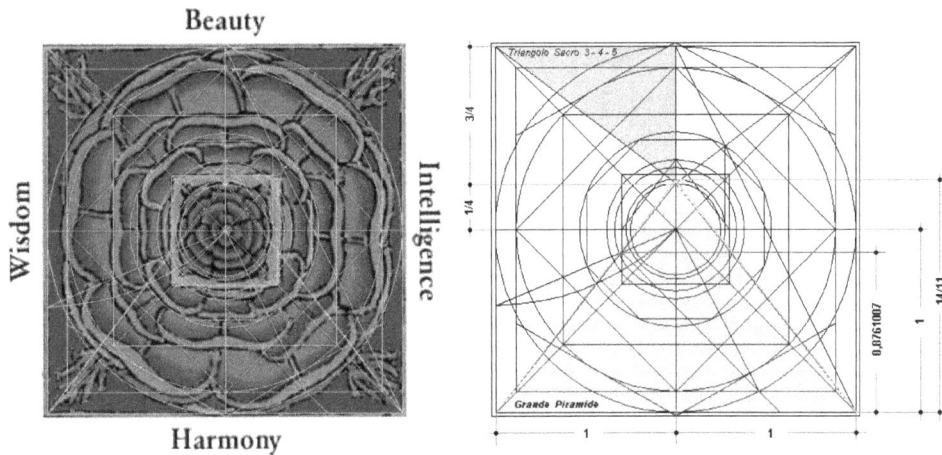

Connie Johnston ● 2003 ● Alfonso Rubino

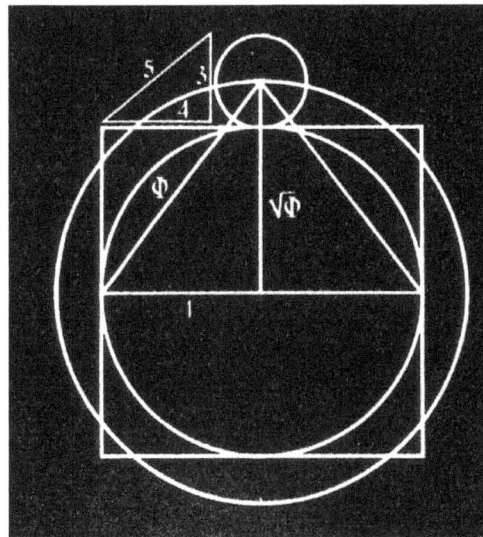

Fig 3
X-Ray View of the Egyptian Pyramid at Gizeh, an anointed shape that makes visible the invisible geometries of Light, (see Fig 4 also), since it is the only currently known shape in the universe that contains both Phi and Pi. The Phi Cone of Fig 1 is based upon this internal geometry. If the apothem radius is 1 unit, the height of the Pyramid is therefore Root Phi 1.272... and the slope height of the Pyramid, not the corner edge length, is Phi or 1.618...

THE GOLDEN RIGHT ANGLED TRIANGLE
Internal View of the Gizeh Square-Based Pyramid

Statistics:
AB = side of Pyramid = 2 units
OC = Apothem = 1 unit
OD = Height = Root Phi = 1.272... units
DC = Slope height = Phi = 1.618033...
DB = Edge Length, is different to DC

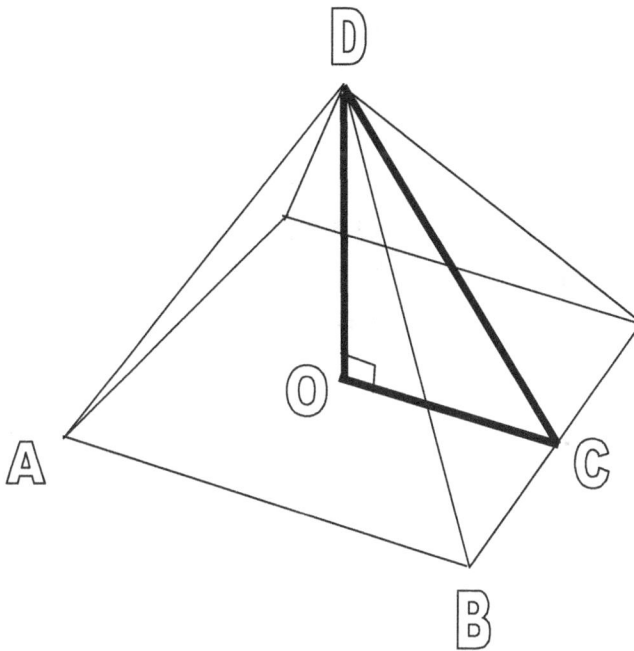

Fig 4
**The Golden Right Angled Triangle based on Unity, Phi and
the Square Root of Phi shown as the guts of the pyramid.**

Orgone Pyramid with quartz crystal point. Based upon the research of Wilhelm Reich, orgone energy transmutes negative energies in surrounding areas. Used for healing noxious earth energies.

Large size measures 5 tall; small 4 tall

Orgone Pyramid

Fig 5
The Phi Cone as designed by Wilhelm Reich
Structured as an Orgone Pyramid
mounted with a quartz crystal at its apex.

Jain 9-1-2010 Mullumbimby Creek (MCk)

APPENDIX PART 8:

WHY 24
(chapter excerpts taken from my Dictionary of Numbers known as **THE HARMONIC STAIRWAY**)

● Why did the **Ancient of Days** refer to their **Pantheon of 24 Deities** as being presided over by **The Invisible One**? Why did they consist of 12 Demigods + 12 Goddesses? (taken from CXXXV1 Manly P Hall's book, The Secret Teachings Of All Ages).
Is this Divine Duodecimo referring to the occult significance of 24-ness when we view 12-fold Deity as Male and Female? This is the same as the 24 Elders which is the Dual Hierarchs (2x12) or Elohim working with the Chromosomes of Man-Woman.
12 Tribes, 12 Prophets, 12 Patriarchs etc.

● Why was it also mentioned in the ancient Gnostic text: "The Pistis Sophia" (by JJ Hurtak), referencing to what Jesus said to his disciples: **"I have come forth from that First Mystery, namely the 24th..."** and this riddle can only be solved by looking at the gematria (sacred number) of the Alpha and the Omega; the number for Jesus being 8-8-8. Nb: **8+8+8 = 24** and 37x24= 888
Nb by Jain: notice the occurrence of the 888 in the 6th column of the Phi Code 1 when it is arranged as a 3x8 array:
It is copied here again: (taken from Chapter 2, Fig 7 on page 80 of this book):

1	1	2	3	5	8	4	3
7	1	8	9	8	8	7	6
4	1	5	6	2	8	1	9

Fig 7
3x8 Rectangular Array of Phi Code 1 showing repetition in the 2nd and 6th columns of 111 and 888, highlighted.

● Why did **King Arthur** have a Round Table made for **24 Knights**?

● Why did the Masons refer to the **24 inch rule**?

● Why does the clock have **24 hours**? Is it our choice for the Division of the Circle to represent Time?

• Why did the Rabbis hold sacred the **24 Threads** or separate strands woven in each twisted linen thread used in weaving the Tabernacle curtains and ornamentations? (According to artist and scholar **Lily Moses** of Mullumbimby, the 24 threads or strands represent our **DNA**!).

• The Phi Code is a **StarGate**. The **Pineal Gland** is a StarGate. **DMT** (DiMethylTrypatamine is a chemical cousin Penta-Hexa Molecule to Seratonin, and since the Pineal Gland is covered with a crystalline structure composed of micro-crystalline calcite that secrets DMT) is a portal, a door for the transport of the Spirit out of the Body. This is the Energy signature of 108, teaches us how to navigate, how to steer the tornado from the Eye of the Spiral, non-destructively, in a self-organized passageway.

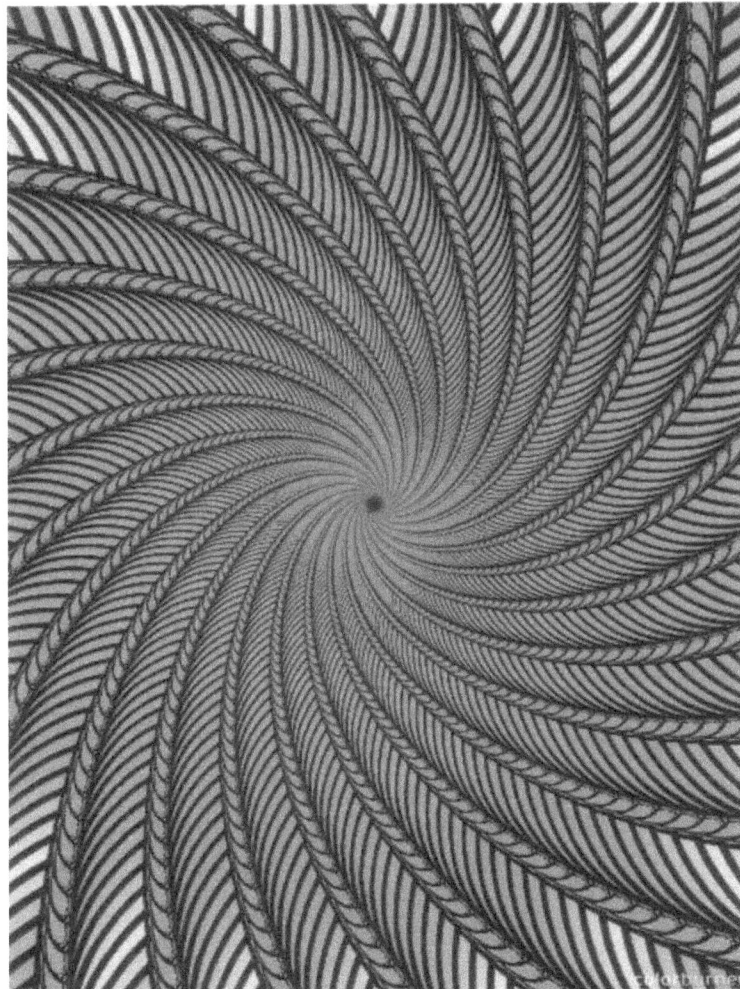

APPENDIX PART 9:
PHI TO 20,000 DECIMAL PLACES
Taken from the Golden Number Website

```
1.61803398874989484820458683436563811772030917980576286213544862270526046281890
2449707207204189391137484754088075386891752126633862223536931793180060766726635
4433389086595939582905638322661319928290267880675208766892501711696207032221043
2162695486262963136144381497587012203408058879544547492461856953648644492410443
2077134494704955846788509874339442212544877066478091588460749988712400765217
0575179788341662562494075890697040002812104762177111778053153171410117046665
9914669798731761356006708748071013179523689427521948435305678300228785699782977
8347845878228911097625003026961561700250464338243776486102838312683303724292
6752631165339247316711121158818638513316203840052221657912866752946549068113171
5993432359734949850904094762132229810172610705961164562990981629055520852479035
2406020172799747175342777592778625619432082750513121815628551222480939471234145
17022373580577278616008688382952304592647878017889921990270776903895321968198615
14378031499741106926088674296226757560523172777520353613939362107673893764556060
60592165894667595519004005559089502295309423124823552121241544400647034056573479
76639723949499465845788730396230903750339938562102423690251386804145779956981224
4574717803417312645322041639723213404444948730231541767689375210306873788034417
00939544096279558986787232095124268935573097045095956844017555198819218020640529
05518934947592600734852282101088194644544222318891319294689622002301443770269923
00780308526118075451928877050210968424936271359251876077788466583615023891349333
1223105339232136243192632728910670503399282265263556209029798642472597772565508
6154875435748264718144512700060238901620777322449943530889990950168032811219432
04819643876758633147985719113978153978074761507722117508269458639320456520989698
55567814106968372884058746103378105444390943683583581381311168993855576975484149
14453415091295407005019477548616307542264172939468036731980586183391832859913039
60720144559504497792120761247856459161608370594987860069701894098864007644361709
33417270919143365013715766011480381430626238051432117348151005590134561011800790
50638142152709308588092875703450507808145458819906336129827981411745339273120809
28972792221329806429494687824274874017450504067787570832373109759151177629784428
47479081765180977872684161176325038612112914368343767023503711116330725869883258
71033632223810980901211019899176841491751233134015273384383723450093478604979294
59915822012581045982309255287212413704361491020547185549611808764265765110604545
88147560443178479858453973128630162544876114852021706440411166076695059775783257
03951108782308271064789390211156910392768384538633332156582965977310343603232254
57436372041244064088267375843395367959312322134373209957498894699565647360072959
99839128810319742631251797141432012311279551894778172691415891177991956481255800
18455065632952859859100090862180297756378925999164994642819302229355234667475932
695165421402109136301819472270789012208728736170734864999815625547281137347987165
6952748900814438405327483781378246669174442296349147081570073525457070897726754
69343822619546861533120953357923801460927351021011919021836067509730895752895774
68142295433943854931553396303807291691758461014609950550648036793041472365720398
60007355076090231731250161320484358364817704848181099160244252327167219018933459
63786087875287017393593030133590112371023917126590470263494028307668767436386513
27106280323174069317334482343564531850581353108549733350759966778412445696074128
30860240332456395357212524261170278028656043234942837301725574405837278267996031
739364013287627701243679831144643694767053127249241047167001382478312865650649343
41803900410178053395058772458665575522939158239708417729833728231152569260929959
42240000560626678674357923972454084817651973436265268944888552720274778747335983
53672776140759171205132693448375299164998093602461784426757277679001919190703805
22046123482391326104327191684512306023627893545432461769975753689041763650254785
1382463146583363833760235778992672988632161858395903639981384582764491245980937
04305559613797834134830494946968610893512569636348428178128862536460368420339465
38194419457142666823718394918323709085748502665680398974406621053603064002608171
12665995419936873160945722888109207788277203636684481532561728411769097926665523
84688311371852991921631905201568631222820715599876468423552059285371757807656050
36773130975191223973887224682580715974457404842987807352215984266766257807706201
94304005425501583125030175340941171910192989038447250332988024501436796844169479
59545304591031381162187045679978663661746059570003445970113525181346006565535203
47888117414994217486204152135567793940390710387088182338068033500380468001748082
20591096844202644640201877053401003180288166441530913939481564031928227854824145
10503188825189970074862287942155895742820216657062188090578805032467699129728721
03870736974064356674589202586656573978560859566534107035997832044633634648548949
76638853510455272982422906998488536968280464597457626510
```

434359050938321243743333870516657149005907105670248879858043718151261004403814
880407252440616429022478227152724112085065788838712493635106806365166743222327
767755797399270376231914704732395512060705503992088442603708790843334261838413
597078164829553714321961189503797714630007555975379570355227144931913217255644
012830918050450089921870512118606933573153895935079030073672702331416532042340
155374144268715405511647961143323024854404094069114561398730260395182816803448
252543267385759005604320245372719291248645813334416985299391357478698957986439
498023047116967157362283912018127312916589952759919220318372356827279385637331
265479985912463275030060592567454979435088119295056854932593553187291418011364
12187470752628106869801357605247194455932195535961045283031488391176930119658
583431442489489856558425083410942950277197583352244291257364938075417113739243
760143506829878493271299751228688196049835775158771780410697131966753477194792
263651901633977128473907933611119140899830560336106098717178305543540356089529
290818464143713929437813560482038947912574507707557510300242072662900180904229
342494259060666141332287226980690145994511995478016399151412612525728280664331
261657469388195106442167387180001100421848302580916543383749236411838885646851
43150063731904295148146942431460895254707203740556691306922099804819452975110
65046428105417755259095187131888359147659960413179602094153085853323877253802
327276329773721431279682167162344211832018028814127474431688472184593927814354
740999990722332030592629766112383279833169882539312620065037028844782866694044
730794710476125586583752986236250099982323597155072338383324408152577819336426
263043302658958170800451278873115935587747217256494700051636672577153920984095
032745112153687300912199629522765913163709396860727134269262315475330437993316
581107369643142171979434056391551210810813626268885697480680601169189417502722
987415869917914534994624441940121978586013736608286907223651477139126874209665
137875620591854328888341742920901563133283193575622089713765630978501563154982
456445865424792935722828750608481453351352181729587932991171003247622205219464
510536245051298843087134443950724426735146286179918323364598369637632722575691
597239543830520866474742381511079273494936523964792689939083689324917999502789500
060459661313463363024949951480805329017902975182515875049007435187983511836032
722772601717404535571658855578297291061958193517105548257930709100576358699019
297217995168731175563144485648100220014254540554292734588371160209947945720823
780436871894480563689182580244499631878342027491015335791072733625328906933474
123802222011626277119308544850295419132004009998655666517756640953656197897818
380451030356510131589458902871861086905893947136801484570018366495647203294334
374298946427412551435905843484091954870152361403173913903616440198455051049121
169792001201999605069949664030350863639209341007019450532016234872763232732449
439630480890554251379723314751852070910250636859816795304818100739424531700238
804759834323450414258431406361272109602282423378228090279765960777108493915174
887316877713522390091171173509186006546200990249758527792542781659703834950580
106261553336910937846597710529750223173074121778344189411845965861029801877874
274456386696612772450384586052641510304089825777754474115332076407588167751497
553804711629667771005876646159549677692705496239398570925507027406997814084312
496536307186653371806058742242598165307052573834541577054292162998114917508611
311765773172095165656478695474489271320608063545779462414531066983742113798168
96382353330447788316933972872891810366408269856988254438516675862289930696434
684897514840879039647604203610206021717394470263487633654393195229077383616738
981178124248365578105034169451563626043003665743108476654877780128577923645418
522447236171374229255841593135612866371670328072171553392646325730673063910854
10886808574283588280602303341408550390973538726134511962926415995212789311354
431460152730902553827104325966226743903745563612286139078319433570590038148700
89866131539819538574423304419708558966722293142730741384882788975588860799738704
470203166834856941990965480298249319817657926829855629723010682777235162740783
807431877827318211919695280051608791572128826337968231272562870001500182929757
729993579094919640763442861575713544427898383040454702710194580042582021202344
580630345033658147218549203679989972935353919681213319516537974539911149424445
183033858841290401817818821376006659284941367754317451605409387110368715211640
405821934471204482775960541694864539878326269548013915019038995931306703186616
706637196402569286713887146631189192685682691995276457997718278759460961617218
868109454651578869122410609814197263644091926315359472922825080542516906906
8140107817960218853307623055638163164019224540325765739259976517530801427160
714308718862859836037465057134204670083432754230277047793311183666903232885306
87387907135900740304907459889513647687608678443238248218930617570319563803230
819719363567274196438726258706154330729637038127515170406005057594882723856345
156390526577104264594760405569509598408889037620799566388017861855915944111725
09231327977113803294376547509016516949650991607833937715833230245701948347400
070437618671998483401631826008462619656284649118225688857521346375490254180833
82138352224525872678937950537591560357945498509102256225455003017571049469833
483545323835260787092219304581782306012370753280678368541306584636788866433486
249368010198782799630670259543265137806007386392908564830874157618741897345848
450141889765293411013722158643559915527113623322003526677859159890231446163321
026519665907632061524383747619049531582968836265042094840105654589130629827717

```
24980964195947234046511041982134768935401803825695495628603924426415986748598 2
28006035386283916620125282660749330619658496519997941939322601723571073364253 7
08303301143362498575363597042444647599899995085550413549775585859345765909265 33
30725277541675843146693676780617035012003844874883823376034407751594778122188 3
07090008738662736209166079905022698927032189976037950989059108591039296734561 4
61070030845819212738925992696106211676436424383501410204086321499178152979681 52
23798322427375365700855346997965541385905032683616022278847554706269843910885 2
10302076860470680455684656049168649886061622295232390709809262930233795648217 9
98163264582788877674520846371971063478923106675469355047615197781699025881840
40792751090182448278705250597698375351430622445090220238243982312550584162320 7
18831930069360646468209659500654929010971618652636721610741713618377667332797 5
62685480124565768279031760394655539452314338756773034979157858859101166374845 5
67584795271391860878254010423332985744274711896961048512640197504359909207662 1
55899866073683762318835884508129295011466535482817144846405686524654090781547 1
61962578446957526256945516560151916402921798854890937328031465192247590030965
71549050536104377686877261915952844920464786897347370859841384513162119297201 2
63424077369454598186502965923353451256845497454112981973587667072860161605620 4
23063606613028149677344579773775055756466547525632264817711699785708712283154 3
10456912326250349768115245217449739613674882204648051968875434196951193312045 0
21605142938484475452382127014383095785581361967830231068508084587695205905329 4
68338490471209916255636503400343967082893698367423001575117385151269123066172
27641442160751291734187471431509324192491416096999867281582385925735982389484 9
27491964615227227333874631213843626211637946706203263022505548958057308375046 1
29923113629917306948940734258831948399927416395098443963405763528471756276219 2
78652253960872013108004864065343961688754525342630989695176190197709631922587 09
34216595597447175015753837674152228057065028068314335652491719973335840306415 3
55075911597426436648284662813680217450590970589460274429263222221545945075804 6
57120606863990430823693969320823749076756119017156130542481331171524256847846 3
36377001520441791650116823257526316040957479063902244344451035121904881980302 76
00176680985096524543900719909803499302686067552387968529219473239335237008665 0
22140746455403722234348167574937314464092837900653919677401035586193618156683 6
61686489239555496145282647289499416061580304586789146197172815545110005666054 2
49969197410279874059327643495371452516769462069859788094695017473022841427571 8
87194092120913799405943037050436483860043464522799330292390186592268987499211 3
25656055784014233542605895105620369072028939315920440476835927636479960059640 4
86076198915929819495087878602766345990540426377004590080327943472062982544525 6
35647954299248819864613617131446570364995334755771554913842392894017540341399 73
84616948129347924223460974301962752301382860722449638095383840152656781976450 7
58854785515549234523478164603306293884200995080326014091830257438577067102522 7
24366690598890854501557075423031666592472352892470258862479488754625276572728 5
15111287827067345431024451523345654228431103967952829625019369893998347396176 3
98809573541526014537296468147382184360052109947211941659149471670520379225520 9
63364584846804144778030216472862399926404836350877374782450163820089524032253 4
37992579012926564015553775409175170441962728503912669595666487242967660367303
45366873404907914188694521471582790817233969124039958869390855170379801955546
12851340891206108401221361707057043006056924685591646883477332085689141267942 8
44804138468281325692914816010978627269686686737391711893146226913489458042778 9
89960814470952476290501926031164920686774331866154696689660182266357878875060 8
85624356267893279735463390418210877463803921624477202567269959639182468778845 5
49717903851583920474831990312762243706623509251877543141010711233586590774812 2
06376345901988422547272765529050439950252444039113658267081330058058820946031 0
20826134136912757293699289302996173089284367031523898937538937388936807441526 373
79424050644876417176861355234326986572897046306918017427797217388985944328485 2
05725758833756382015054672065167425268189485167332804630764781329313260289322 9
36604521021318981298766152624448748669389040617846991666541748508459797014617 8
21584501491957210982508923451747451225432738681972586494458808377139868506598 4
08545773165416917406705211194916628633773226375347566637002212032752438999773 6
00607404270297220363477804829883485518952507947460551994034011077116972564426 1
00509205984336253584706959718576261677663021174787834197564450183804102920324 0
40882661734433909026352235050682858285443283961848092537613082011562686990799 9
11708475586982150310073563240421988569584200682439926953784403202222374628147
65923060554747693683057654967769047115962550247450780962483744990802561375091 5
62235908101053449394177429427709144516666870041522854463807661535114155648785 4
93601138747310382877331338839170964617482906315678806518276176579835021665998
60746401267488412113009854993833710603196250670279752431011937733554853701169 4
67485888836308033328773957165627534036727218070562256232637414883349928997025 8
97729922403694175074342731419415743246679457858603989407509735636368881567215 9
67635438066559393893438207598406121609463176644219026777737991455799450314687 08
71626622652413359056992849400637274490882163524294802256633045855363633725176 2
04907462406293896239062203042487268843237763173357420575399757437350840965779 2
18088008942059066257278230769278865644556375801266728095252737982803007663697 6
92816484465127747382239706173856750714669274822037488112256399407522762646499 4
```

6584636740195599737028383931198848223355399649783331650084674912545229565124093909637840954169012346753752801390808308630226533523870692730719846546494549791011342871546366955434374621543918865260853669743665305885621644116480689128373577943415306094784572709870379769213462059695388438267608276591817736276699187278037542199541724283357910645206137368847085451658221931586453770183134018188272510999229176147118605291765514228811235662172416926806206488453176151642729535857983754123758761004154758055957301224592767118952773338233560433742013213928043170533794636464283519930145767064918477077689598854216479733717696259439386480748936332010988936435283244941325693174383235092582864212762094734328799843871982916250358863688574408960916197675530236361478401862718277088913603989330772930602967177602584180301334754744060932182226620770598424760826379413885986019352089598219418857238237142719304935451824011267104607309741268127907272643868568154472914482676138994509206409879264769257469881233464299526730823740572040614374870086704861259959017842497684584473682482794782475317633817481479957103120339634522674341512372232245462654632835356424662778646083987217912784308964163642223715282219986085060015824516947831892606016582749114277493350286550372769106810755782646334039921922260220859096784186001385965387726582624465759769406924054180444473847160790144974301805588933762376129691822923476845375955564684211226987316375062499711822914856896044725277600939343435583391951651329856236458931491018608496834803380909327362610620547959704212986698835735604043471283998012498022094668510934904078784501021176842763450791376876097469006657596830435192666765639609226488456702128507448211848361029076891964934023006417531734839147589166720230692453471076277197925249973285768903886801417803137994836510895272209465913045066566582585391746904868726499025467659665991645473651342597555773973485065284399773844905139058294301300083669614556697485377934078812772157914872107192588690892778873298298221457423327326598798275695089884530624022303648634772296705652412703588783028194007498057543901628578674553132719765260710764315311239152607721936214434609608975872693422367433161371871857457760811775151806966210479558514013490407878450102117684276345079137687609746900554957627391245587148332010170361840521636818017341425089806160634676330850504184585816629334093479199103685913053789482158651701181210113330006695775232786685518078256752836149494920745837336845813691407975959252672739664234787466143998196480810367050660052382691650551446347111168674281773195025606429516379596594756449878914614469259366293093648048161740598082142543405252113713324081139135799716228581014191034104605692907824989562145600410456922214168308932366625176186962717194538549985514842751733692412026801599280832014583007544847423312643878084780850561043049099993643459051951874948436967727574733596708833496091574474357503986020163976661142765369526704411552001939148429346010151295311744588764830703716773961542655913990830375776630213099087127198870690329304701241058615063998529981417578043034808035882032020110476070047557101694234120341089156439478253031645937304375581946867525349532301302767823535601166413111779960997936620434495696835479307543113275586431897315151710644321892497932778012649647644754670781658074061312593752718474088161154798183078167510478092914139545646311605812690517539535569157755804106719812316384052775560522722237647118832332230995850689710187175047819065334948584232597622565758418985291447178335173226029857862929434650563669321626276738162459574179326988923272206666360819924909888314685299409913867344604967084244297824363023293891035596560173994220198869025724547140163300961214618720836510868818533406062201709951582707044233704218017669634913369599606432200532887349489313596603042438080456594474333567831672703729636367594216999379522

INDEX + NUMBER HARMONICS

Nb: any page numbers <u>underlined</u> means that this reference is an **image** or a **chart**, not text.

HARMONICS INDEX: NUMERICAL DICTIONARY OF NUMBER REFERENCES:

NB: Any reference to an image or chart is underlined.

THE BOOK OF PHI, Volume 4
The 108 Codes: The Linear Phi Code 1

As a sequel to the Da Vinci Code, already based in the mass consciousness, here is the next riddle, solved.

Jain 108, the mathematical sleuth, the synesthete, the philomorph (lover of forms), the **mathematical monk**, the **numerical nomad**, the para-physical code-breaker, has gone deeper down this Phi Code 108 rabbit hole. In volume 3, he introduced us to the Phi Code 108 **pine cone maths**, and now in this epochal book of volume 4, he reveals extraordinary phine details never seen before in print.

This 108 compendium is like a mathematical detective story having a denouement as magical as it is indisputable, as baffling to professors of mathematics, geometry and technophiles as it is simple in its precise and grand design being lucidly based on nature's rawest eco and bio principles.
To even consider than an elegant, symmetrical and infinitely repeating 24 code-pattern exists in the **Living Curvation** of **Nature's Numbers,** will be hotly denounced and widely read.
For 2,000 years we have been duped that no pattern exists in the Divine Proportion. How sad that this distortion was ever believed!
This 108 Code, **Pine Code**, this pulse of the Universe, is the stuff of cryptically coded parchments, buried treasures, secret societies and obscure dynasties of oppression and suppression of knowledge.
What is all the fuss about:

1-1-2-3-5-8-4-3-7-1-8-**9**-8-8-7-6-4-1-5-6-2-8-1-**9**

How can it be that over a billion Hindu people in the world are worshipping this **Shri 108** yet they do not know its mathematical derivation or origin!

Jain 108's Digitally Compressed and Infinitely Repeating 24 Pattern is indeed another permanent bullet in the world of fortified mathematical armouring.

Jain's central problem or "spiritual core issue of concern" is that he is a man of the 22nd Century living in the 21st!
Poetical and Passionate, he is attempting to bring in the feminized, right-brain mathematics based on 108 and the True Value of Pi, but is challenged appropriately and vigorously by rigid curriculum-makers who fear change and a loss of job. He is an intrepid Mathematical Psychonaut and **Mathematical FUTURIST**.

ISBN: 978-0-9757484-3-5

www.ingramcontent.com/pod-product-compliance
Lightning Source LLC
Chambersburg PA
CBHW081525220326
41598CB00036B/6332